Genetics, Genomics and Breeding of Maize

Genetics, Genomics and Breeding of Maize

Editor

Sangeeta Karnik

scitus academics

Genetics, Genomics and Breeding of Maize

Edited by **Sangeeta Karnik**

Printed in 2017

ISBN: 978-1-68117-023-7

Library of Congress Control Number: 2015931525

SCITUS Academics LLC,
616, Corporate Way, Suite 2, 4766,
Valley Cottage, NY 10989

www.scitusacademics.com

Contents

Preface

Genomics is defined as the study of genes and their functions, and related techniques. The main difference between genomics and genetics is that genetics scrutinizes the functioning and composition of the single gene where as genomics addresses all genes and their inter relationships in order to identify their combined influence on the growth and development of the organism.

The plant genetics and genomics addresses the fundamental nature of genes and genomes at any level, e.g., from molecular to chromosomal and from population genetics to comparative genomics, by means of either experimental or computational approaches, in plants as well as algae, including applications of genetic and genomic knowledge to plant biology and plant improvement.

Maize or corn is a cereal crop that is grown widely throughout the world in a range of agroecological environments. More maize is produced annually than any other grain. About 50 species exist and consist of different colors, textures and grain shapes and sizes. White, yellow and red are the most common types. The white and yellow varieties are preferred by most people depending on the region.

Maize is the most important cereal crop in sub-Saharan Africa (SSA) and an important staple food for more than 1.2 billion people in SSA and Latin America. All parts of the crop can be used for food and non-food products. In industrialized countries, maize is largely used as livestock feed and as a raw material for industrial products.

Sequencing of the maize genome has opened up new opportunities in maize breeding, genetics and genomics research. This book highlights modern trends in development of hybrids, analysis of genetic diversity, molecular breeding, comparative and functional genomics, epigenomics and proteomics in maize.

Editor

Chapter 1

Advances in Maize Genomics and Their Value for Enhancing Genetic Gains from Breeding

Yunbi Xu,[1] Debra J. Skinner,[1] Huixia Wu,[1] Natalia Palacios-Rojas,[1] Jose Luis Araus,[1] Jianbing Yan,[1] Shibin Gao,[1,2] Marilyn L. Warburton,[1,3] and Jonathan H. Crouch[1]

[1]International Maize and Wheat Improvement Center (CIMMYT), Apdo. Postal 6-64, 06600 Mexico, DF, Mexico

[2]Maize Research Institute, Sichuan Agricultural University, Ya'an, Sichuan 625014, China

[3]USDA-ARS-CHPRRU, Box 9555, Mississippi State, MS 39762, USA

ABSTRACT

Maize is an important crop for food, feed, forage, and fuel across tropical and temperate areas of the world. Diversity studies at genetic, molecular, and functional levels have revealed that, tropical maize germplasm, landraces, and wild relatives harbor a significantly wider range of genetic variation. Among all types of markers, SNP markers are increasingly the marker-of-choice for all genomics applications in maize breeding. Genetic mapping has been developed through conventional linkage mapping and more recently through linkage disequilibrium-based association analyses. Maize genome sequencing, initially focused on gene-rich regions, now aims for the availability of complete genome sequence. Conventional insertion mutation-based cloning has been complemented recently by EST- and map-based cloning. Transgenics and nutritional genomics are rapidly advancing fields targeting important agronomic traits including pest resistance and grain quality. Substantial advances have been made in methodologies for genomics-assisted breeding, enhancing progress in yield as well as abiotic and biotic stress resistances. Various genomic databases and informatics tools have been developed, among which MaizeGDB is the most developed and widely used by the maize research community. In the future, more emphasis should be given to the development of tools and strategic germplasm resources for more effective molecular breeding of tropical maize products.

INTRODUCTION

Maize is a widely grown C4 crop with a high rate of photosynthetic activity leading to high grain and biomass yield potential. It is predominantly a cross-pollinating species, a feature that has contributed to its broad morphological variability and geographical adaptability. Depending on the latitude and the climate in which it is grown, maize is classified into three distinct types, tropical, temperate, and subtropical. Maize can also be classified based on: (1) endosperm and kernel constitution; (2) kernel colour: flint, dent, floury, waxy, sweet, and popcorn; (3) maturity; (4) use. Economically, the most important types of maize are grown for grain or fodder and silage production. However, in the tropics, grain is primarily grown for human consumption. FAO predicts that an additional 60 Mt of maize grain will be needed from the annual global harvest by 2030. The demand for maize as an animal feed will continue to grow faster than the demand for its use as a human food, particularly in Asia, where a doubling of production is expected from the present level of 165 Mt to almost 400 Mt in 2030 [1].

In addition to being an economically important crop, maize is also a classical genetic model for plant research. It has a number of characteristics that are favorable for an experimental model for crop plants: (i) a multiple-purpose crop with worldwide cultivation which attracts research funding from public and private institutions, (ii) 59 000 [2] and 42 000–56 000 genes [3] with moderate genome size (2400 Mb of DNA per haploid nucleus in the B73 inbred, which is approximately six times larger than rice and six times smaller than wheat, although a large proportion of the genome represented by repetitive elements), (iii) outbreeding reproduction system with tolerance of inbreeding, (iv) existence of multiple breeding products (inbreds, hybrids, synthetic cultivars, open-pollinated varieties (OPVs), improved landraces), and (v) wide adaptability including good sources of resistance to environment stresses.

The objective of this paper is to overview various aspects of maize genomics, including genetic and molecular diversity, genetic mapping and trait tagging, physical mapping and genome sequencing, functional genomics, transgenics, nutritional genomics, genomic databases and tools, and genomics-assisted plant breeding. Throughout the paper we have attempted to synthesize the current status in these areas with a particular reference to implications for impacts on application and molecular breeding. Finally, we provide a brief outlook on future developments in this field and the resultant opportunities that they may provide for the development of new maize products.

GENETIC, MOLECULAR, AND FUNCTIONAL DIVERSITY

The maize genome harbors tremendous molecular diversity that mirrors its substantial phenotypic variability. When considering nucleotide polymorphism in genes, two maize lines are on average as diverged from one another as humans are from chimpanzees [4]. Understanding the useful genetic diversity for crop improvement should speed the development of new, more productive and better adapted cultivars.

Molecular Genetic Diversity

Molecular markers have been extensively used in maize genetic diversity studies for: (1) analysis of genotype frequencies for identification of deviations at individual loci [5] and for characterization of molecular variation within or between populations [5–9], (2) construction of "phylogenetic" trees [5, 7, 10–14] and determination of heterotic groups [7, 13, 14], and (3) analysis of correlation between genetic distance and hybrid performance, heterosis (when the hybrid shows vigor superior to its parents), and specific combining ability [11, 15].

Several studies have demonstrated a decline in genetic diversity across the elite temperate maize pool over the past century [16, 17]. This supports earlier conclusions that landraces and wild related species contain untapped sources of biotic and abiotic stress resistance that should provide useful new alleles for future maize improvement [18]. After analyzing over one hundred maize inbred lines and teosinte accessions with 462 SSRs, Vigouroux et al. [19] concluded that many alleles in the progenitor species of maize (teosinte) are not present in maize. Wright et al. [20] compared single nucleotide polymorphism (SNP) diversity between maize and teosinte in 774 genes and concluded that maize accessions had a far lower genetic diversity consistent with products of artificial selection and crop improvement.

There is a growing awareness that levels and patterns of allelic diversity within specific chromosomal regions provide an important chromosomal context for each locus in that region. "Diversity maps" showing the distribution(s) of allelic diversity across chromosomes and genomes of a variety of organisms suggest that there is an association between chromosomal structural features (such as centromeres and telomeres), and selection in particular well-defined gene pools [21–23]. In addition, diversity analysis of individual genes is shedding new light on crop productivity and evolutionary processes underlying plant domestication [4, 24–26]. Maize is a crop that has high-resolution genetic maps and, therefore, it is an ideal choice for development of diversity maps that could provide new information about the consequences of natural selection, domestication, and polyploidy formation. Correlating variation at the molecular level to phenotypic diversity is an essential prerequisite for future studies of diversity using the large number of available candidate genes. Quantitative trait locus (QTL) information can then be combined with association approaches to select a small number of candidate genes that have a high probability of being directly related to a specific phenotype.

Molecular Marker-Based Core Collections and Allele Mining

Molecular genetic diversity analysis has provided a powerful tool to enable efficient and systematic sampling of the diverse material from breeding programs and germplasm collections. Data from molecular markers have been used to construct core subsets, which maximize the diversity of the original collection in the minimum number of genotypes [27, 28]. The Generation Challenge Program (GCP) has been coordinating the identification of minicomposite germplasm collections for around twenty of the most important staple crops in developing countries. These subset collections are now being intensively characterized at the molecular and phenotypic levels to find new functional diversity for important agronomic traits. Meanwhile, maize is being intensively screened via resequencing and precision phenotyping to test the feasibility of the whole "population" based approach for simply inherited traits. For example, using association mapping and allele mining to identify optimum SNP haplotypes for use in molecular breeding programs [29].

Molecular and functional diversity of the maize genome can be characterized through allele mining, identification of distinct "haplotypes" for different inbred lines, analysis of single feature polymorphisms (SFPs), discovery of nearly identical paralogs (NIPs [30]), and determination of their evolutionary implications. In general, there are two approaches that have been elaborated for allele mining: resequencing and ecotilling [31]. Ecotilling is not being widely used in maize at this time, due to the very high number of sequence differences found between different maize accessions, which confound the interpretation of the differences in such a complex system. Nevertheless, genomewide genotyping using gene-based markers can be used as the foundation of the resequencing method for allele mining in diverse germplasm. However, this approach suffers from the fundamental challenge of establishing which of the various alleles present is functionally different from the wild type, and identifying which new alleles beneficially influence the target trait. Methods to ascertain allele function include marker-assisted backcrossing (MABC), genetic transformation, transient expression assays, and association analysis using an independent set of germplasm from that used to identify the allele. As more genetic variation is functionally validated, it is hoped that the growing database of comparisons between sequence variation and phenotype will allow bioinformaticians to identify patterns that can form the basis of future predictive models. Meanwhile, the rate limiting factor for the effective use of outputs from allele mining in breeding programs continues to be insufficient information on the relationship between SNP variation and changes in phenotypes that may be useful for breeders. However, resources and tools necessary to perform in silico trait targeted selection of the outputs from allele mining are becoming available. Thus, proof-of-concept projects are now being carried out in model organisms, in order to study the relationships between SNP haplotypes and changes in phenotypes. This has already led to the development of predictive tools that can identify those SNPs with a high probability of conferring deleterious phenotypes. However, the next big step in this area is the development of bioinformatics tools to compare sequence variation with protein and functional domain variation or with public databases including associated phenotype data, in order to predict which subselections of SNP haplotype variants have the maximum likelihood of providing beneficial phenotypic variation in the target trait. It is likely that SNPs in promoter and noncoding regions will also be important for predictive phenotype analysis.

The same methodology used in association mapping may also be used for allele mining of the diverse core subsets of maize created from breeder's lines, genebank accessions, and wild relatives. Once a gene of interest is positively identified (via association mapping or any other technique) and the sequence determined, this gene can then be resequenced (entirely or in part) in all the individuals of the subset. Changes in the DNA sequence, corresponding to new alleles of this locus, will be identified in this manner, and individuals carrying the new alleles can be evaluated for the target trait to determine the associated change in phenotype and the value for subsequent use in breeding programs. These alleles may never have been found via simple phenotypic screens, either because it is not possible to grow and measure every plant in a large germplasm collection under all possible environmental conditions, because their effect may be masked in certain genetic backgrounds, or because the effect may be so small that it will not be found unless specifically sought in carefully controlled phenotypic screens.

Sequence validated SFPs between maize inbred lines can be detected by hybridizing RNA or complexity-reduced genomic DNA to an Affymetrix GeneChip expression array. Gore et al. [32] evaluated the efficacy of four different complexity-reduction methods for sensitive SFP detection in maize: cDNA, methyl filtration, high-C_ot selection, and AFLP. These four methods were applied to three diverse maize inbred lines (B73, Mo17, and CML69) with three replications per line (36 GeneChips). The results indicate that all genome reduction methodologies offer modest power to detect SFPs with the commercially available GeneChip Maize Genome Array.

The cost of sequencing is steadily reducing and sequencing increasing numbers of maize genotypes is becoming increasingly possible. Thus, diversity studies at a functional level based on sequencing can be used to complement (and reveal more fundamental variation) studies using genetic and molecular marker analysis. In a recent example, a massively parallel pyro-sequencing technology commercialized by 454 Life Sciences Corporation was used to sequence the transcriptomes of shoot apical meristems isolated from two inbred lines of maize (B73 and Mo17) using laser capture microdissection (LCM) [33]. Putative SNPs were computationally identified, with over 36 000 putative SNPs detected within 9980 unique B73 genomic anchor sequences. Stringent postprocessing reduced this number to around 7000 putative SNPs. Over 85% (94 of 110) of a sample of these putative SNPs were successfully validated by Sanger sequencing. Based on this validation rate, this pilot experiment conservatively identified more than 4900 valid SNPs within over 2400 maize genes. These results demonstrated that 454-based transcriptome sequencing is a method of choice for the high-throughput acquisition of gene-associated SNPs.

GENETIC MAPPING AND GENE TAGGING

The efficiency and precision of genetic mapping and gene tagging has significantly improved in recent years particularly due to the development of sequence-based markers and array-based genotyping systems. Both private and public sectors have established array-based genotyping systems for SNPs, which makes high-throughput genome-wide genetic mapping possible. A large collaborative effort to determine and define SNP loci in genes has been combined with a highly multiplexed array-based genotyping system for genetic analysis of maize [34, 35]. For example, this can be based on the Illumina Golden Gate assay system which facilitates the simultaneous assay of large numbers of SNP loci (e.g., 1536 as currently available) in an arrayed format.

Linkage Mapping

Genetic mapping in maize was first carried out using morphological markers generating a genetic map consisting of 62 morphological trait loci [36]. The first generation of molecular marker maps in maize was constructed using restriction fragment length polymorphisms (RFLP) [37, 38], which were later saturated with simple sequence repeat (SSR) and other types of PCR-based markers [39]. Most recently, linkage mapping is being raised to a new level as maps are being developed with large numbers of SNP markers and/or candidate gene-based markers.

Linkage mapping, mainly using RFLPs and SSRs, has been carried out in maize by numerous laboratories since the 1980s and large amounts of data are now available in the MaizeGDB database (http://www.maizegdb.org). Several types of mapping population have been used, including F_2 [37, 40], immortalized F_2 [41, 42], and populations of recombinant inbred lines (RILs) [38, 43, 44]. These mapping efforts have involved up to 214 individuals screened with 92 to 1736 markers. Composite maps have also been constructed from multiple crosses [45]. In order to improve the resolution and extend the total map distance, the Maize Mapping Project (MMP) developed RILs through several generations of intermating an F_2 population derived from the single cross of the inbred lines B73 and Mo17. As a result, the resolution of the genetic map was improved significantly, consisting of about 1000 RFLP and 1000 SSR markers (see [39]; MaizeGDB, http://www.maizegdb.org). A second panel of intermated RILs (IRILs) developed from F2 F252 was used for linkage mapping of 1454 maize candidate genes [46], which created a higher map resolution with more cDNA loci mapped than when using nonintermated RILs. The maize sequencing project (using inbred line B73) and the constant progress in maize functional genomics are providing new genes and functional genomic DNA sequence information that are increasingly being integrated into the maize genetic map [47]. A total of 25 908 markers have now been integrated into the fingerprinted BAC contig (FPC) map [48]. This includes 1902 genetically mapped markers (SSRs, RFLPs, SNPs, and InDels) and 24 006 sequence-based markers (ESTs, BAC ends, and 40-bp overlapping oligonucleotide overgo probes) [49]. Compared to other types of molecular markers, SNP markers have several advantages, including high abundance and even distribution through the genome. In addition, SNP markers provide highly reproducible codominant information, and there is an increasing range of cost effective high-throughput SNP genotyping systems. It would be expected that the second generation of molecular markers such as SSRs will soon be replaced by third generation SNP markers in genomics research and genomics-assisted breeding. At the International Maize and Wheat Improvement Center (CIMMYT), over 2000 SNP markers that were developed for SNP chip-based genotyping through collaboration with Cornell University and Illumina Inc are being mapped using three RIL populations.

Gene/QTL Mapping

Molecular marker-facilitated mapping of genes underlying specific traits in maize was first reported in the late 1980s by Stuber et al. [50] followed by Edwards et al. [51] and Stuber et al. [52]. Since then, more than 2000 QTL related to various traits of agronomic importance in maize, including yield, yield components, plant morphology and physiology, and biotic and abiotic stress responses have been reported (http://www.maizegdb.org). Two examples which address special issues related to typical quantitative variation should be mentioned here. In one study, two elite inbred lines were crossed to create a 1000-individual mapping population which was evaluated in 19 environments for grain yield, grain moisture and plant height [53]. In the second example, the high and low oil and protein content lines derived from 70 generations of long-term selection were crossed, intermated, and mapped [54]. Both studies reported numerous QTL of very small effect, supporting the concept that quantitative variation is the product of numerous minor genes. This hypothesis is also supported by long-term large-scale QTL studies on drought tolerance in maize that have found no "major effect QTL" [55]. However, as highlighted by Buckler et al. [4], in all these cases, the parental genotypes of the mapping populations probably tapped only a small proportion of the functional variation available in the maize genepool. Plus some important QTL for complex traits may be fixed in the advanced breeding material used as parental genotypes in some of these studies.

Most reported QTL have been identified using bi-parental mapping populations. This approach suffers from the problem that the estimated position of the QTL can vary significantly if different genetic backgrounds are used. This is particularly troublesome for more complex traits such as yield, yield components and abiotic stress tolerance. In addition, these traits have low heritability and high genotype-by-environment interaction (GEI). Hence, these QTL results may only be applicable to breeding populations closely related to the mapping population and that are targeted for same environment conditions as used for the evaluation of the mapping population. Thus, it is important to provide more generally applicable information for the development of marker-assisted selection (MAS) systems and to expand the utilization of QTL results in a range of breeding populations. This may be achieved by surveying the overlapping regions or colocalization of QTL for the same trait across different mapping populations and diverse growing conditions using a core set of markers to generate a consensus map. Bioinformatic tools have been developed to facilitate integration, comparison, and collective analysis of such data. For example, the Comparative Map and Trait Viewer (CMTV) was developed to integrate data collected by CIMMYT in maize drought tolerance research with data from public sources [56]. In addition, QTL for grain yield, kernel weight, abiotic response [57], and disease [58] have been integrated into a consensus map based on data available at the web-based MaizeGDB database. A comprehensive review of all QTL studies in maize is beyond the scope of this paper. However, Tuberosa and Salvi [59] have summarized the main results from the large number of studies that have described QTL for grain yield and other agronomic traits in maize.

Two methods that can save time and money when attempting to identify markers that tag QTL are bulked segregant analysis (BSA) and selective genotyping (reviewed in [60]), where only selected individuals representing the two phenotypic extremes of the target trait are genotyped. BSA is based on bulking the DNA of the selected plants while selective genotyping focuses on individual plants. Selective genotyping can be used to replace entire population genotyping if a sufficient population size is used. This has the advantage of focusing on the individuals that contribute the most useful data to the analysis while then being able to increase the number of markers applied [61, 62]. These two methods are particularly useful for identification of large-effect or major QTL. BSA has been successfully applied to detect major QTL in maize for yield and leaf abscisic acid (ABA) concentration under drought conditions [63], which were not easy to detect with traditional mapping populations. To develop maize molecular breeding systems, CIMMYT scientists are now using large-scale selective genotyping to test whether elite alleles for drought tolerance, disease resistance and grain quality traits can be identified among large numbers of carefully chosen breeding lines that are genotyped with a large number of SNP markers.

QTL mapping has also been used to study the domestication process for cultivated maize. Domesticated maize and its wild progenitor, teosinte, differ dramatically in their overall plant architecture and the morphology of their female inflorescences. Five major QTL have been identified that differentiate maize from teosinte [64]. Two QTL conferring major morphological differences were defined as single Mendelian loci: teosinte branched1 (tb1) [24, 65, 66] and teosinte glume architecture 1 (tga1) [67, 68]. The allele tb1 suppresses lateral branching (leading to apical dominance) while tga1 affects the hardness of the seed coat (hard casing that envelops the seed in its ancestor teosinte), and both genes were involved in the evolution of cultivated maize from teosinte. More recently, additional key loci controlling the differences between maize and teosinte have been identified through QTL analyses of maize-teosinte intercross progeny using a total of 1723 progeny genotyped with more than 300 SNP, SSR, and candidate gene markers and phenotyped for 22 morphological traits [69]. In another report, the regions in the maize genome that affect tassel branch number were identified [70].

Association or LD Mapping

Association mapping or linkage disequilibrium (LD) mapping has become an increasingly important tool for gene mapping in maize [71–80]. LD often declines to within 1 to 1.5 Kb in maize landraces and a broad range of tropical and temperate inbred lines [81, 82], but with less rapid decay in elite breeding materials [83]. Association mapping in plants can be based on candidate genes or whole genome scanning. The latter has become increasingly applicable in maize due to the recent development of large numbers of SNP markers.

Comprehensive association mapping studies are being carried out in the Buckler group at Cornell University (http://www.maizegenetics.net/), where sequence variation in thousands of genes with important functions is being studied across a diverse panel of 288 maize genotypes. CIMMYT is currently applying association mapping techniques in order to identify markers for provitamin A accumulation in colored maize kernels. CIMMYT is also using these approaches for more complex traits such as drought tolerance, using drought tolerance related candidate genes and phenotyping over 300 diverse maize genotypes in multiple locations over two years.

To facilitate genome-wide association mapping of both qualitative and quantitative traits, an integrated mapping strategy, Nested Association Mapping (NAM), was designed consisting of 25 maize populations, each of which has 200 RILs derived from crossing one of 25 diverse inbred lines to a common inbred [84]. With a dense coverage (2.6 cM) of common-parent-specific (CPS) markers, the genome information for 5000 RILs can be inferred based on the parental genome information. Essentially, the linkage information was captured by the CPS markers and the LD information among loci residing between CPS markers was then projected to RILs based on parental information, ultimately allowing for genome-wide high-resolution mapping. The power of this approach using 5000 RILs allows 30% to 79% of the simulated QTL to be precisely identified [84, 85]. Preliminary analysis of flowering time has led to the identification of 50 QTL significant beyond the

LOD 4 threshold. Although there was a 40-day variation in flowering time between the tested lines, no allele with more than 3-day effect on flowering time was observed and the vast majority of the allelic effects provided just 0.5–1.0 day changes in flowering time. Next steps include QTL identification, examination of the recombination events in specific populations, and association analysis of SNP datasets. This will lead to the identification of single genes or candidate genes underlying key QTL controlling the target traits (Ed Buckler, personal communications). Combining the complete genome sequence with NAM approach will greatly facilitate the dissection of complex traits in many species able to apply this strategy. At CIMMYT, a strategy has been developed to integrate linkage- and LD-based mapping. This is being used for mapping drought tolerance using two 1536-SNP chips, one of which was developed from candidate genes related to drought tolerance [86]. Compared to linkage-based mapping, the advantages of LD-based mapping are: no need for generation of mapping populations; high mapping resolution; mapping of multiple traits using the same set of germplasm. However, the power of association mapping is variable. Although whole genome scan mapping of traits has been validated in maize, in general, we do not have a detailed understanding of the power of this approach, given the highly variable LD, noisy phenotypic data and complex population structure [87].

Validation of Marker-Trait Associations

A wide range of approaches to validation of marker-trait associations has been reported, each with their own profile of advantages and disadvantages. The availability of thousands of SNP markers, rather than hundreds of SSR markers, makes it practical to validate marker-trait associations through high-precision genotyping of an independent set of parental lines and breeding populations. Marker validation can also be carried out by simultaneously mapping multiple populations or by selective genotyping of phenotypic extremes from multiple relevant breeding populations. A similar level of validation may be achieved when applying association mapping with a large number of diverse inbred lines. Greater efficiency in the validation process can be achieved through pooled DNA analysis provided that markers have been selected and optimized for this approach. However, validation requirements may be minimized when dealing with large-effect QTL

[88]. Alternatively, the "mapping-as-you-go" approach provides the opportunity to validate and/or refine the marker-trait associations at every generation [89]. It is likely that approaches based on the use of breeding materials for mapping will become increasingly popular due to the overall time and cost efficiency. At the same time, it is likely that the utilization of haplotype-based selection rather than single-marker-based selection will become increasingly common in view of the increased selective power that it provides, particularly for complex traits and/or multiple trait selection.

An alternative strategy for confirming candidate QTL is the use of near-isogenic lines (NILs). Reducing much of the "noise" caused by the effects of genetic background, NILs offer much more accurate estimates of QTL effects than RILs, particularly if multiple QTL are segregating in the population. However, the power to detect an individual QTL may be greater in RILs than NILs [90, 91]. For example, a major L-ABA (leaf-abscisic acid concentration) QTL in bin 2.04 affecting root traits and relative water content was further confirmed using NILs [63]. Grain yield [92] and flowering time [91] traits have also been mapped using this method.

WHOLE GENOME SEQUENCING

Two large-scale efforts in the USA were launched during the early 2000's to sequence the maize genome. The Sequencing the Maize Genome Project (STMG) was a collaboration between the Plant Genome Initiative at Rutgers University (J. Messing), and the Arizona Genomics Institute and Computational Laboratory (R. Wing and C. Soderlund). Working from the high-resolution FPC genome sequencing began by producing BAC-end sequences (BES, also known as Sequence Tag Connectors [STC]) [2]. These sequences help to order the BAC's as well as provide novel sequence and a repeat sequence database. Shotgun-sequenced BACs provide valuable new sequence and also important data on genome structure [3].

Given the structure of the maize genome, strategies for targeting gene rich regions have been the focus of the second major maize sequencing initiative. The Consortium for Maize Genomics (CMG), consisting of the Donald Danforth Plant Science Center, The Institute for Genomic Research (TIGR), Purdue University and Orion Genomics, has used selective techniques for enriching for genes. Two strategies, methyl-filtration and high-selection, were used to enrich for gene rich regions [93, 94]. Methyl filtration is based on the finding that the genic regions of maize are not methylated, and can be selected accordingly. High- selection follows the rate of rehybridization of genomic fragments following denaturation, with repetitive sequences reannealing quicker than low copy number genic sequences. Both strategies have been shown to be very effective, capturing 95% of genes in test BACs, and are likely to be more efficient than random whole genome shotgun sequencing alone [95, 96]. Assembled Zea mays sequences from these techniques can be viewed at TIGR (http://maize.tigr.org/release4.0/assembly.shtml). In addition, the MAGIdb contains Maize Assembled Genomic Islands (MAGIs) using almost four billion genome survey sequences generated through the CMG which can be viewed and used to BLAST against other sequences (see [97]; http://magi.plantgenomics.iastate.edu/) with basic local alignment search tool (BLAST).

In addition to these efforts, other methylation based methods developed for maize will help to extend genic "islands" and also order these islands relative to each other. Hypomethylated partial restriction (HMPR) libraries generated by methylation sensitive restriction enzymes can be larger than the methyl filtration clones as they are tolerant of internal methylation sites [98]. Therefore, they can cover the gene island and extend the available sequence for any particular gene. Another approach, methylation spanning linker libraries (MSLLs) also uses methylation sensitive enzymes but selects for large fragments, where the ends are anchored in neighboring genes, indicating linkage, orientation, and distance between genes [99].

A full-scale program to sequence the entire maize genome was initiated in 2005 through the NSF-funded Maize Genome Project as a collaboration between the Washington University Genome Sequencing Center, the Arizona Genomics Institute (AGI), Iowa State University, and Cold Spring Harbor Laboratory, aiming to sequence the maize genespace of the cultivar B73 to a finished quality using a BAC-based approach (NSF award DBI-0527192). The strategy is to utilize a minimal tiling path of approximately 19,000 mapped BAC clones to focus on producing high-quality sequence coverage of all identifiable gene-containing regions of the maize genome. These regions will be ordered, oriented and, along with all of the intergenic sequences, anchored to the Agarose FPC physical map and the IBM genetic maps of the maize genome (maizesequence.org). Development of maize physical map of the B73 inbred is essential for the use of the sequence information. In February, 2008, at the 50th Maize Genetic Meeting, the Maize Sequencing Consortium announced the draft sequence of the maize genome. Since then, extensive work has been done to finish the remaining clones, improve the physical map, anchor the sequence to the genetic map, build an AGP (A Golden Path) to generate maize pseudomolecules, and to annotate the genome. On March 20, 2009, AGI released an updated maize integrated genetic and physical map, consisting of 440 contigs.

A recent report (ScienceDaily, June 28, 2008) indicates that a major maize sequencing project is being carried out at the Center for Research and Advanced Studies of the National Polytechnic Institute (CINVESTAV, Mexico), using bulked plants from the Mexican popcorn landrace Palomero. This maize accession has 22% less DNA and is phylogenetically closer to teosinte than B73. The project has been focusing on important gene rich regions (http://www.niherst.gov.tt/s-and-t/s-and-t-news/; Dr. Alfredo Herrera Estrella, CINVESTAV, personal communications). Structural and functional analysis of this genome reveals a large number of unreported sequences, suggesting that the ancient landraces contain a large pool of unexplored genetic diversity that could be useful in new crop generation as well as the study of the evolution and domestication of maize and other cereals (http://abstracts.aspb.org/pb2008/public/S02/S024.html).

With next-generation DNA sequencing technology [100, 101], we may be able to sequence any maize genotype of interest. A new generation of high-throughput sequencing technologies promises to transform the scientific enterprise, potentially supplanting array-based technologies and opening up many new possibilities [102]. Significantly enhanced sequencing throughput will allow us to uncover the huge diversity of novel genes that are currently inaccessible, to better integrate biological information for a complete picture of many traits at an individual level and to move to advances that we cannot yet imagine.

FUNCTIONAL GENOMICS

Functional genomics seeks to determine gene functions and interactions using genomic scale data. An important part of new gene discovery through EST and genome sequencing is the annotation of those genes to assign putative functions. In the absence of empirical data for a particular gene in the appropriate target organism, gene annotation software can predict a function using data from rice, Arabidopsis and other organisms based on similarities for intron/exon sequence and structure plus likely protein domains (http://www.maizesequence.org). Users can BLAST their own maize sequences against the cereal and other species databases to generate predicted function information. In forward and reverse genetics approaches, maize is characterized by excellent mutagenesis resources in the form of well-studied transposon systems, and new techniques for induced mutations are also being applied [103]. The completion of the maize genome sequence provides the most essential resource to move easily from gene to mutant phenotype and back. There are several methods for experimentally determining gene function. Only three important fields will be discussed.

Gene Cloning

Major challenges for gene discovery include: (1) the large size of the maize genome; (2) variation in genome size and gene order; (3) the high incidence of multicopy genes; and (4) transposons and other repetitive sequences make up a large portion of the genome. Traditionally, gene discovery in maize has employed transposon tagging, EST searches and comparative genomics, and due to the increase in genomic resources, positional cloning is increasingly being used for both qualitative and quantitative traits.

Positional cloning in Arabidopsis and rice has been very successful because of their small genomes and the availability of the complete genome sequence. Until recently, positional cloning in maize was considered nearly impossible because of the vast amounts of repetitive DNA. However, now with the physical maps for maize, the large numbers of available markers, and critically, conservation of synteny across the cereal genomes, it is feasible to consider a chromosome walk in comparable or less time than cloning by transposon tagging [104]. The basic steps are the development of a large segregating population, of at least 1000 individuals, and initial mapping using a subset of the population with around two markers per chromosome arms to generate an approximate map position of the gene of interest. The remainder of the population is then genotyped with the flanking markers to identify recombinants from the whole population that are suitable for the fine mapping process. Additional markers are then screened across these recombinants, and where markers from maize are not available, syntenic rice or sorghum markers can be used. These methods should allow determining the physical position of the target gene at the level of an individual BAC contig which equates to an average distance of 1 cM [104]. However, this depends on highly precise phenotyping, and is more rapid for qualitative traits whose phenotypes are easily determined. To assist the process, outputs from the maize genome project (including overgos, anchoring of ESTs, and BES) and the fully sequenced genome of rice can be used to identify candidate genes. These genes can then be sequenced to look for the mutation and in the process could provide additional markers to narrow down the location of the target gene. As more genomics resources become available, including the genome sequence of maize and its relatives, positional cloning of quantitative traits will be most limited by the need for precision phenotyping of the trait under study.

An important and surprising finding from comparing sequence data from Mo17 and B73 is that on average 50% of the sequence at a genomic locus can differ between these two inbred lines [105]. This might be expected to significantly interfere with the positional cloning process. However, the majority of sequence differences between inbreds are caused by insertion or deletion of partial genes or pseudogenes carried by Heliotron transposons. The genes that are collinear between inbreds are usually collinear with rice, meaning that if the rice genome is used as a reference for positional cloning it is unlikely to be affected by the differences between maize inbreds. Where the genomic differences between inbreds are significant, more work will be needed to find useful polymorphisms, but given the extensive diversity between maize lines, these polymorphisms are likely to be abundant. On the other hand, however, it is not uncommon to see genes that are contiguous in rice being split across two regions in maize [106, 107]—this can make cloning difficult. The first genes identified by positional cloning in maize were QTL as these were more difficult to target with transposons. However, this type of trait presents a serious challenge due to the difficulty of accurately screening the phenotype and the need to work with large population sizes. A maize domestication locus, teosinte glume architecture (tga1), which encodes a transcriptional regulator, was the first maize gene positionally cloned, using a population of over 3000 individuals [108]. A QTL for flowering time was localized to a conserved 2 kb noncoding region using a combination of positional cloning, comparative genomics, and association (linkage disequilibrium) mapping. The noncoding region was shown to act as an enhancer of a distant flowering gene [77].

Positional cloning was recently used to identify an RNA polymerase encoding gene (rmr6) which influences paramutation and maize development [109]. In this case, the predicted maize genes within the mapped interval were absent from the syntenic interval in rice, but four of these predicted genes formed a single large gene which was homologous to an Arabidopsis polymerase gene. Several genes that regulate inflorescence architecture in maize have also been cloned. The gene responsible for the mutated phenotype of a highly branched tassel and a branched ear, ramosa 1 (ra1), was cloned by transposon tagging [110]. Two other maize inflorescence genes have been cloned using the map position of the mutation in combination with prior candidate gene information. Thebarren stalk 1 (ba1) mutants lack tassel branches and spikelets and are missing ears. The positional cloning of lax panicle in rice [111] provided a candidate gene for ba1. In a second example, a maize clavata 1 (clv1) ortholog was mapped to chromosome 5 in the same region as thick tassel dwarf 1 (td1). The phenotype of td1 mimics that ofArabidopsis clv mutants, which have larger inflorescence meristems and more floral organs. Proof that td1 was the clv1 ortholog came from analysis of a large number of Mu-induced alleles [112]. In addition, ra2, ra3,tasselseed4 (ts4), and sparse inflorescence1 have also been cloned using the positional approach described above [113–115]. Cloning of indeterminate gametophyte 1 used a combination of positional cloning and transposon insertion to validate candidate genes [116].

Silage maize is an important source of forage for dairy cattle due to its high energy content and good digestibility. Brown midrib (bm) mutants in maize have an increased digestibility but inferior agronomic performance [117]. Two of the four bm genes (bm1 and bm3) have been shown to be involved in monolignol biosynthesis [118]. These two and additional lignin biosynthesis genes have been isolated based on sequence homology with rice. Candidate genes putatively affecting forage quality have been identified by expression profiling using isogenic bmlines, and associations detected between a polymorphism at the COMT (caffeic acid O-methyl transferase) locus and DNDF (digestible neutral detergent fiber) in a collection of maize inbred lines [119, 120]. Conserved domains or motifs shared by across known resistance genes have been extensively exploited to identify unknown resistance gene analogs (RGA). In an attempt to isolate all potential RGA based on these domains, three approaches were adopted by Xiao et al. [121], including the modified AFLP (amplified fragment length polymorphism), modified RACE (rapid amplification of cDNA ends), and data-mining methods. In response to herbivorous insects, plants synthesize and release volatile chemical signals that will attract the natural enemies of the herbivore to defend themselves. A recent report described the findings that the function of the maize Hm1resistance gene may be conserved in other grasses to protect these plants against the lethal fungal necrotrophic pathogen Cochliobolus carbonum [122]. Lin et al. [123] reported the isolation and characterization of the maize monoterpene synthase gene tps26 that is an ortholog of stc1, a gene induced in response to the attack of beet army worm larvae. Maize orthologs of the tomato pto-interacting (pti) protein have been identified through homology and library screening [124]. The tomato gene functions in signal transduction of the defense response toPseudomonas syringae, while at least one of the four ZmPti1 genes functions in pollen signaling. Examples of functionally characterized genes in maize include Dwarf8 that encodes a gibberellin response modulator from which functional markers can be developed for plant height and flowering time. For example, nine sequence motifs in the Dwarf8 gene of maize were shown to be associated with variation in flowering time, and one 6 bp deletion accounted for 7–11 days difference in flowering time between inbreds [71]. However, Dwarf8 is a pleotropic gene (also affecting plant height) and thus there is a need to identify functional markers from additional flowering time genes for use in breeding applications. Recently, a streamlined PCR-based cloning strategy and massively parallel sequencing are used to clone and characterize the expression of a maize COBRA-like gene required for cell wall expansion in root hairs [125]. A list of additional characterized maize genes is maintained at MaizeGDB (http://www.maizegdb.org/newgenes.php).

Transposon Tagging and TILLING

The wealth of active transposable elements residing in the maize genome plays an important role in functional genomics. In addition to serving as molecular tags for mutated genes, these transposons tend to knock out genes into which they insert. Mutant libraries constructed using transposon tagging, T-DNA insertion, and chemical and physical mutagenesis provide materials that can be screened for base changes in any genes by reverse genetics methods such as Targeting Induced Limited Lesions IN Genomes (TILLINGs).

Large populations of maize plants containing highly active Mutator transposons have been created to saturate the genome with insertional mutations, and these form the basis for different transposon tagging resources (reviewed in [103]). May et al. [126] described an efficient system for site-selected transposon mutagenesis in maize. A total of 43 776 F_1 plants were generated by using Robertson's Mutator (Mu) pollen parents and self-pollinated to establish a library of transposon-mutagenized seed. The frequency of new seed mutants was between 10^{-4} and 10^{-5} per F_1 plant. As a service to the maize community, insertions in genes of interest from this library can be selected by using the PCR. A maize-targeted mutagenesis database was established for storing pedigree, knockout, sequence, phenotype, and other information.

McCarty et al. [127] implemented a novel strategy for harnessing the power of high-copy transposons for functional analysis of the maize genome, and reported behavioral features of the Mutator system in a uniform inbred background. The established unique UniformMu population and database facilitate high-throughput molecular analysis of Mu-tagged mutants and gene knockouts. The Mu populations differed markedly in the occurrence of Mu insertion hotspots and the frequency of suppressible mutations. The public database (http://uniformmu.org; http://endosperm.info) contains pedigree and phenotypic data for over 2000 independent seed mutants selected from a population of 31 548 F_2 lines and integrated with analyses of 34 255 MuTAIL sequences.

Chemical and radiation-induced mutations have been widely used for random mutagenesis in plants, resulting in a broader spectrum of mutation alleles that occurs randomly in the genome. A number of chemicals have been used to generate large mutant collections, many of which are described in MaizeGDB and available for study. Chemical agents generate a broader range of DNA alternations; these are predominantly single base-pair substitution, but also induce small insertions and deletions. Ethyl methane sulfonate (EMS), a base-alkylating agent that generates point mutations (of which the vast majority are G/C-A/T transitions, often leading to the creation of stop codons/nonsense mutations), has been used most commonly because of its ease of use and the diversity of potential mutants.

Strategies have been developed so that subtle changes like point mutation generated by EMS can be detected easily. For efficient adaptation of chemical induced mutagenesis for reverse genetics in Arabidopsis and other plants, McCallum et al. [128] developed the TILLING screening system, which allows a point mutation to be identified at a large-scale. In the basic TILLING method, seeds are mutagenized by treatment with EMS. These mutagenized lines serve as a general forward genetic resource. EcoTILLING [31], as a variant of TILLING, examines natural genetic variation in populations and has been successfully utilized in animals and plants to discover SNPs. The protocols developed for TILLING have been adapted in EcoTILLING for discovery of natural nucleotide variation linked to important phenotypic traits. Till et al. [129] reviewed the current TILLING and EcoTILLING technologies and discussed the process that has been made in applying these methods to many different plant species. In maize, the Maize TILLING Project (MTP) started doing TILLING screens in 2005 using mutant populations created in different laboratories (http://genome.purdue.edu/maizetilling/). Weil and Monde [130] provided a detailed protocol for maize TILLING including TILLING mutagenesis, tissue collection, DNA preparation and 2D pooling, and detailed TILLING workflows. As the maize genome is completely sequenced, advances in reverse genetics technologies including TILLING, EcoTILLING, and massively parallel DNA resequencing provide excellent methods for identifying mutations in a wide variety of traits and biological processes [131].

TRANSCRIPTION PROFILING

To provide comprehensive, low-cost, and public sector long-oligonucleotide (around 70 mers) microarrays for gene expression analysis in maize, the first generation of low cost oligonucleotide spotted microarrays was developed for the maize community [132]. A total of 57 452 70 mer oligonucleotides were designed to represent 25 969 ESTs assemblies, 20 206 singleton EST (detected only in a single cDNA library), 9707 assembled maize sequences, 804 nonredundant repeat elements, 467 organelles, 288 maize community favorites and 11 transgenes. Replicated baseline expression profiles have been generated for 18 tissues and deposited in Zeamage as have the results of other expression studies (http://www.maizearray.org). The GeneChip Maize Genome Array manufactured by Affymetrix (http://www.affymetrix.com) is a commercial alternative to the public 70 mer array described above. This array contains 17 555 probe sets, which interrogate 14 850 maize transcripts representing 13 339 maize genes. These arrays have 25 mer probes, with 15 different probes designed from the end of each gene. A recent advance includes whole-genome transcript profiling with a 100 K Maize Affymetrix GeneChip Array, which contains 100 000 probe sets to detect transcripts from Zea mays.

Transcript abundance levels differing between the parental genotypes of a mapping population and segregating among the progeny can be mapped and characterized as quantitative traits [133]. Microarrays have been used to determine the gene expression levels and identify genomic regions (gene expression QTL, or eQTL) associated with transcript variation in coregulated genes. The eQTL mapping involves expression profiling as measured by mRNA transcript abundance for a large number of genes which are each treated as a quantitative phenotype likely to be conveyed by multiple genes and influenced by environmental factors. These expressional profiles then constitute a marker-based fingerprint of each individual in a segregating population and can be subjected to conventional QTL analysis [134] albeit interpreted in the spatially and temporally specific context in which the data were collected.

As one of several combined approaches, gene expression profiling has been used to study the molecular basis of heterosis in maize. In a recent eQTL report, all possible modes of gene action, including additivity, high- and low-parent dominance, underdominance, and overdominance, were observed in a comparison of global gene expression in a maize F_1 hybrid compared with its inbred parents, B73 and Mo17 [135]. Over one thousand genes were identified as being significantly differentially expressed between the three genotypes. In a second report, microarray analysis of gene expression patterns in immature ears, seedlings, and embryo tissues from the inbreds B73 and Mo17 identified numerous genes with variable expression. The reciprocal F_1 hybrid lines did not display maternal or paternal effects on gene expression levels. The results suggest that cis-transcriptional variation between B73 and Mo17 led to additive expression patterns in the F_1 hybrid [136]. This group has also reported other properties of gene expression in some maize hybrids [137–139]. In a third report, a wide-scale survey of transcriptional heterosis in maize immature ears has been carried out using B73 and H99 inbred lines and the resultant F_1 using cDNA microarray technology and real-time PCR (RT-PCR) [140]. Genes expressed at a significantly different level between parental genotypes and the F_1 hybrid were identified. Both dominance and over dominance components were reported to be involved in nonadditive gene expression variation in the studied ear developmental stage, encompassing a wide variety of biological processes. Other examples of nonadditive gene expression have been described in a root transcriptome study, which identified a gene that was consistently expressed above the midparental value for different parent and hybrid combinations that may function in heterosis [141]. In addition, massively parallel signature sequencing (MPSS) [142], a deep sequencing-based mRNA profiling technology, was used to study allele specific expression. This enabled a genome-wide evaluation of cis- and transeffects on allelic expression in six meristem stages of the maize hybrid [143].

Gene expression profiling has been widely used in the study of abiotic stress tolerance (e.g., [144] as reviewed in [145–147]). In maize, flowering is the developmental stage most vulnerable to abiotic stress leading to significant yield loss associated with aberrant floral development and impaired ear and kernel growth. Genes within the starch biosynthetic pathway are collectively downregulated under stress leading to reduction in starch content. Comparative profiling of the sense and antisense transcriptome revealed that transciptomes of the three lines tested (A619, ND101/W23 and W23) displayed remarkable similarities across four tissues (leaves, 1 mm anthers, 1.5 mm anthers, pollen) despite high levels of polymorphism and structural differences between the inbred lines [148]. Transcriptome analysis of the low-phosphorus responses in roots and shoots of a phosphorus-efficient maize line identified alterations of several metabolical and physiological processes [149]. Maize seedlings were surveyed for transcription changes under six abiotic stresses, including the agronomically relevant treatments cold and dessication [150]. Overlapping expression profiles and coordinate expression indicated genes relevant to stress resistance, and such datasets form an excellent resource for identifying candidate genes through positional cloning or association mapping.

Transcription profiling has increasingly become an important genomics tool for gene functional analysis. Research results that have been published so far revealed great variation in gene expression for a large number of genes. However, further functional analyses are needed to understand how these genes contribute individually and together to a specific differential phenotype. As numerous QTL have been located in chromosomal regions using molecular markers and the interaction effects between these have been studied, transcription profiling may follow a similar process beginning with discovery of large numbers of eQTL and then focusing on ones with large effect for more detailed functional analyses and analysis of interaction effects. Before their function can be concluded, it needs to be confirmed that all putative eQTL have a specific influence on the target trait.

TRANSGENICS

Genetic transformation in major cereal crops has become a powerful research tool for gene validation, as well as enabling the introduction of novel genes directly into breeding pool and thus accelerating or complementing conventional breeding efforts. In maize genetic transformation has been extensively used in crop improvement, particularly for the development of new commercial pest and herbicide resistant cultivars but more recently also including more complex traits such as grain quality and drought tolerance.

Transformation Methodology

The past ten years have witnessed extensive efforts toward the development of an efficient Agrobacterium-mediated transformation system for maize with particular emphasis on increasing the efficiency and extending the range of amenable genotypes [151]. Although the biolistic approach (physically shooting DNA into cells) has revolutionized the genetic transformation field for major crop species, it is usually associated with high copy numbers, transgene silencing and rearrangement. Agrobaterium-mediated transformation approach (biologically facilitated entry of DNA) is believed to generate a high proportion of independent events with single or low transgene copy numbers which is expected to favor consistent transgene expression in progeny generations [152,153]. However, since A. tumefaciens naturally infects only dicotyledonous plants, monocotyledonous plants remained inaccessible for many years despite enormous efforts worldwide. The breakthrough was made in 1994 when Hiei et al. [154] reported using A. tumefaciens to transform rice. Two years later, Ishida et al. [155] reported that the successful Agrobacterium-mediated transformation of the maize inbred line A188 and its hybrid using superbinary vectors (cloning vectors, able to replicate in both E. coli and A. tumefaciens. A superbinary vector carries additional virulence genes from a Ti plasmid), and this was confirmed by Negrotto et al. [156] and Zhao et al. [157]. Subsequently, it was demonstrated that maize could also be transformed using A. tumefaciens carrying an ordinary binary vector [158, 159]. For achieving good transformation efficiency, advances have been made through selection of basal media, modifying medium components, optimizing different culture stages, and adding Agrobacterium growth-inhibiting agents such as silver nitrate [157, 159–163]. Yang et al. [164] found that the purine and pyrimidine biosynthesis inhibitors, mizoribine, azaserine, and acivicin could induce higher transformation efficiencies when appropriate concentrations were added before or during inoculation withAgrobacterium. It is believed that this is achieved by inhibiting key enzymes in the de novo purine biosythesis pathway in Agrobacterium cells thus improving the competence of plant cells [165]. In addition, a replication-associated protein (RepA) was used to stimulate cell division and callus growth leading to higher transformation efficiency [166]. More recent developments in transformation technology also include development of novel plasmids and T-DNA binary vectors that incorporate a modified and more useful form of the superpromoter [167] and construction of engineered minichromosomes by modifying natural A and supernumerary B chromosomes [168].

Only a limited number of proprietary [166] or public inbred lines [169–171] and various recalcitrant inbred lines crossed to A188 [161, 172] have been transformed using A. tumefaciens-mediated transformation. Zhao et al. [157] reported using Hi-II which contains A188 and B73 genetic backgrounds, for achieving high-transformation efficiencies up to 40%. Although Hi-II performs very well at the tissue culture and transformation stages, T_0plants have poor seed setting capacity [158]. However, hybrids derived from the crosses between Hi-II and elite germplasm showed many "hybrid vigor (heterosis)" characteristics including more aggressive rooting, thicker stems, and taller stature than plants derived from Hi-II events. The hybrid T_0 plants exhibited excellent tassel development in the glasshouse and the seed set was three to five times higher than Hi-II transformants [173]. Using standard binary vectors, an enhanced Agrobacterium-mediated transformation of Hi-II immature zygotic embryos was achieved recently by employing low-salt media in combined use with antioxidant l-cysteine alone or l-cysteine and dithiothreitol during the Agrobacterium infection stage [159].

Marker-Assisted Breeding for Transformability

Inbred line A188 has been shown to produce highly embryogenic callus in culture, leading to its frequent use in maize transformation investigations into the genetic control of embryogenic response in tissue culture. Early in 1992, Armstrong and Rout [160] proposed that there was a major gene (or genes) in the region marked by probe c595 on the long arm of chromosome 9 highly associated with several measures of in vitro culture response. Two independent mapping studies using A188 reported a total of seven QTL affecting tissue culture and transformability from A188. Willman et al. [174] suggested that at least one gene (or block of genes) controls the expression of the frequency of somatic embryogenesis. To further dissect the genetic basis of embryogenic response in maize, a mapping population of 101 RILs was developed from crossing A188 with a popular (nonembryogenic) maize inbred line, B73. The A188 B73 (BC_3S_5) lines are estimated to contain approximately 3% of the A188 genome and 97% B73 [175]. Six lines were identified that produced a higher than expected number of somatic embryos when they were cultured for two weeks on a regeneration medium containing auxin, cytokinin and abscisic acid. At least one new locus was found that controls the production of somatic embryogenesis in A188. Lowe et al. [176] reported marker-assisted breeding (MAB) for transformability in maize. Maize lines with improved culturability and transformability were produced using MAB to introgress specific regions from the highly transformable hybrid, Hi-II, into the elite line, FBLL that responds very poorly in culture. FBLL is a female inbred parental stiff-stalk line that has been used in some historically best-selling hybrids produced by a seed company DeKalb (now Monsanto). Five unlinked regions important for culturability and transformability were identified by segregation distortion analysis and introgressed into FBLL to produce the highly transformable FBLL-MAB lines.

Genetically Modified Maize

The commercial sector has made substantial progress with pest resistant maize through transformation with genes encoding for insecticidal crystal (Cry) proteins from Bacillus thuringiensis (Bt), which have been particularly successful in providing protection against several corn borers. Transgenic maize plants with the gene encoding snowdrop lectin gene under the control of a phloem-specific promoter were not only resistant to homopterns, but also showed toxicity to Asia corn borer, a type of Lepidoptera [108]. Transgenic plants with thep1 transcription factor resulted in enhanced silk maysin production, thus achieving corn earwarm resistance as elevated concentration of silk maysin causes earworm abiosis [177].

There has been increasing interest in addressing more complex traits such as grain quality and abiotic stress tolerances. Naqvi et al. [178] created elite inbred South African transgenic corn plants in which the levels of three vitamins were increased specifically in the endosperm through simultaneous modification of three separate metabolic pathways. The kernels of the transgenic white corn (Cv. M37W) were found to contain 169-fold the normal amount of -carotene, 6-fold the normal amount of ascorbate, and double the normal amount of folate. More examples for grain quality are discussed in the section on nutritional genomics. For chilling and cold tolerance, Ohta et al. [179] managed to shift the break point C lower than that of the wild type by the introduction of an antisense gene for maize cold tolerant pyruvate orthophosphate dikinase (PPDK) into maize. Low level but constitutive expression of an active tobacco mitogen-activated protein kinase kinase kinase (Nicotiana NPK1) has been found to enhance freezing tolerance in transgenic maize plants that are normally frost sensitive [180]. The gene (NPk1) has been shown to have a significant effect on photosynthetic rates under drought stress when implemented with a modified constitutive promoter 35SC4PPDK [181]. As a consequence transgenic plants produced kernels with weights similar to those generated under well-watered conditions, while kernel weights of drought-stressed nontransgenic control plants were significantly reduced when compared with their nonstressed counterparts [182]. An elite maize inbred line, DH4866, has been transformed with the beta Agene encoding choline dehydrogenase, a key enzyme in the biosynthesis of glycine betaine from choline. The transgenic plants were more tolerant to drought stress than wild-type plants at both germination and young seedling stages. Most importantly, the yield was significantly higher than the wild type [183]. The same group has also produced and analyzed transgenic maize with improved salt tolerance through the introduction of AtNHX 1gene into maize genome, and some lines produced were able to germinate and grow in the presence of 0.8% and 1.0% sodium chloride [184]. Very recently, Castiglioni et al. [185] demonstrated that transgenic maize lines with bacterial RNA chaperones resulted in not only abiotic stress tolerance but also improved grain yield under water-limited conditions. The result supported the hypothesis that the endogenous function of cold shock proteins (CSPs) in plants relies on RNA binding/ chaperone activity through the cold chock domain (CSD) and these proteins, similarly to bacteria, regulates stress responses through a posttranscriptional mechanism. A plant nuclear factor YB subunit (NF-YB) protein in Arabidopsis and an orthologous NF-YB protein from maize were identified coordinating plant responses to drought tolerance. The orthologous maize transcript factor, ZmNF-YB2, was shown to have an equivalent activity. Under water-limited conditions, transgenic maize plants with increased ZmNF-YB2 expression showed tolerance to drought based on the responses of a number of stress related parameters, including chlorophyll content, stomatal conductance, leaf temperature, reduced wilting, and maintenance of photosynthesis. These stress adaptations contributed to a grain yield advantage under water-limited environments. Under relatively severe conditions, the best performing transgenic maize line produced about 50% increase in yield relative to controls in the same experiment. The application of this technology has the potential to significantly impact maize production systems that experience drought [186].

Commercialization of Transgenic Maize and Its Impacts

Transgenic maize has been cultivated commercially in the United States since 1996. By 2000, about 25% of US maize had transgenic resistance to certain insects and/or herbicides, and this proportion increased to about 40% by 2003 [187] and 52% by 2005 [188]. Other countries that had approved releases of genetically modified (GM) maize by 1996 included Argentina and Canada. In Europe, GM maize was approved for use by the governments of Spain and France in 1998. Under European law, any seed which is approved in one EU country is automatically approved in all the others. But the process of extending approval for MON810 (resistance to European corn borer (Ostrinia nubilalis)) beyond France and Spain was suspended for five years by the EU moratorium on new GM products. The moratorium was lifted in May 2004, and the European Commission approved MON810 to be grown in any EU nation. But Greece still refuses to lift the ban on GM maize, while others such as Germany are providing approvals only on a case-by-case basis and primarily for nonfood uses.

Transgenic maizes producing insecticidal toxins from Bacillus thuringiensis (Bt) are widely used to control pests, but their benefits will be lost if pests evolve resistance [189]. The mandated high-dose/refuge strategy (http://www.epa.gov/pesticides/biopesticides/pips/bt_brad.htm) for delaying the development of resistance in the pest requires the planting of refuges of toxin-free crops near Bt crops to promote survival of susceptible pests who may have competitive advantage over resistant pests. To the contrary, resistance to a Bt crop has yet to be documented, suggesting that resistance management strategies have been effective thus far. However, current strategies to delay resistance remain far from ideal [190]. Unfortunately, it is highly difficult for governments in developing countries to enforce the refuge strategy. Moreover, it has been demonstrated that pollen-mediated gene flow (up to 31 m) from Bt maize caused low to moderate Bt toxin levels in kernels of nonBt maize refuge plants, which could seriously undermine the high-dose/refuge strategy and facilitate the accelerated development of pest populations resistance to Bt crops [189]. Thus, it now seems that farmers must implement measures to reduce gene flow between Bt crops and refuge plants. This is likely to be even more difficult than achieving widespread use of the refuge approach. Clearly, there is an urgent need to develop alternative ecologically and evolutionarily appropriate strategies that are easy to deploy in developing countries.

Commercialization of transgenic maize for abiotic stresses such as drought tolerance has been very limited. There are several reasons for this. First, it is very difficult to phenotype the transgenics under either natural or controlled environments. Second, most of the comparisons between wild and transformed genotypes have been performed in conditions that are not very well reflective of developing country cropping environments. This has resulted in the identification and validation of transgene effects on drought tolerance under experimental conditions that are not mirrored under field conditions. This situation is changing now by an ongoing Water Efficient Maize for Africa project (http://www.aatf-africa.org/aatf_projects.php?sublevelone=30&subcat=5), which is a public-private partnership lead by African Agriculture Technology Foundation. In this project, CIMMYT will provide high-yielding maize varieties that are adapted to African conditions and expertise in conventional breeding and testing for drought tolerance. Monsanto will provide proprietary germplasm, advanced breeding tools and expertise, and drought-tolerance transgenes developed in collaboration with BASF.

Quist and Chapela [191, 192] reported the presence of transgenic DNA constructs in native maize landraces that were sampled from northern Oaxaca, Mexico. This raised questions about whether the introduction of large-scale commercial transgenic maize cultivar production would have a deleterious effect on the diversity of maize landraces and traditional agricultural systems of small-scale farmers. An important concern in assessing the risk of growing a genetically modified crop in its center of domestication is gene flow between the transgenic crop and its wild relatives. A second systematic survey of transgenes in currently grown landraces in the state of Oaxaca was carried out using highly sensitive PCR-based markers, appropriate positive and negative controls, and duplicate samples for DNA extraction. No transgenic sequences were observed so the report concluded that transgenic maize seeds were extremely rare (or entirely absent) in the sampled fields [187]. Cleveland et al. [193] analyzed the apparently conflicting conclusions from these two reports and sided with the original report. This was because the samples size used in the second study was not representative, and their statistical analysis was inconclusive due to using census population size c_1 instead of effective population size . Ortiz-García et al. [194] contested this finding and reiterated that there was a clear need for additional surveys with rigorous sampling methods to provide accurate estimates of transgene frequencies over broad geographic areas in Mexico. Large-scale chip-based whole genome fingerprinting, combined with pooled DNA analysis, would provide evidence for solving the conflicts.

NUTRITIONAL GENOMICS

In maize kernels, both macro- and micronutrients are present, including carbohydrates (starch), lipids and proteins (macronutrients), carotenoids, tocopherols, minerals, phytic acid, anthocyanins and other phenolic compounds (micronutrients). Malnutrition has long been recognized as a major public health problem in developing countries, including those where maize is used as staple food. Together with a balanced amino acid profile, attention has also been focused on three micronutrients where nutrient enhancement in maize could contribute to alleviate the problem: iron, vitamin A, and zinc [195]. In addition, the nutritional benefits of phytochemicals such as anthocyanins, xanthophylls, and tocopherols are also well recognized.

The term "nutritional genomics" is used to describe research and product development at the interface of plant biochemistry, genomics, and human nutrition [196]. It also involves the use of metabolic engineering, genetic engineering, and specific technologies. Despite many studies in these areas, there is still a lack of information for some traits, particularly in terms of isolation of regulatory elements and structural genes, duplicate function loci, feedback inhibition, branched pathways, or other phenomena affecting trait expression. Nevertheless, in the following subsections we present some examples of progress in understanding and manipulating phytic acid, iron, zinc, quality protein, and carotenoids.

Phytic Acid, Iron, and Zinc

Phytic acid, known as inositol hexakisphosphate (IP6) or phytate when in salt form, is a virtually ubiquitous component of plant seeds, supplying both phosphate and cations during germination. Phosporous bound in phytate is nutritionally unavailable to monogastric animals and thus contributes to water pollution because it is excreted in the waste [197]. In its native state, phytate forms complexes with proteins as well as mono- and divalent cations, thus decreasing the bioavailability of micronutrients like iron and zinc and exacerbating human mineral deficiencies. Although there are also positive effects against carcinogenesis that have been shown with in vitro cell cultures due to its antioxidant properties [198], phytate is considered an antinutritional compound [199]. Thus, the main advantage of low phytic acid (lpa) mutants is that the bioavailability of various minerals may be improved. However, due to its dual effect on human health, any strategy of reducing kernel phytic acid must consider the needs of the target population.

Several mutants with low levels of phytate have been isolated, and the loci mapped in maize, including lpa 1-1, lpa 2-1, and lpa 241 [200, 201]. Several studies have been conducted to understand phytic acid biosynthesis, identifying genes involved in the pathway. Using Mutator insertion knockout technology, Shi et al. [202] identified a maize inositol phosphate kinase gene involved in phytic acid biosynthesis in developing seeds. The ZmIpk loss-of-function mutant for this gene is allelic to the low-phytic acid mutant lpa2. Cloning and sequencing of the ZmIpk gene from lpa2-2 showed that the lpa2-2 allele has a nucleotide mutation that causes immature termination of the ZmIpk open reading frame. In the lpa2-1 mutant, the genomic sequence was found rearranged in the ZmIpk locus, and no mRNA expression was detected [202]. Despite efforts to elucidate and manipulate phytic acid biosynthesis, low-phytic acid mutants have limited value to breeders because of adverse effects on agronomic traits such as low germination rates, reduced seed weight (lpa1-1), stunted vegetative growth and impaired seed development (lpa241). However, Shi et al. [203] have recently identified the gene disrupted in maizelpa1 mutants as a multidrug resistance-associated protein (MRP) ATP-binding cassette (ABC) transporter. Silencing expression of this transporter using the embryo-specific Glb promoter produced low-phytic acid, high phosphate transgenic maize seeds that germinate normally and do not show any significant reduction in seed dry weight.

To increase the amount of bioavailable iron in maize, Drakakaki et al. [204] have generated transgenic maize plants expressing aspergillus phytase and iron-binding protein ferritin. This strategy has proven effective for increasing iron availability and enhancing its absorption. However, much work is still to be done to transfer this technology to tropical and subtropical maize genotypes normally grown in the areas of greatest need for enhanced iron content maize.

Due to the complex and still poorly understood action of iron and zinc in plant metabolism, not much work has been carried out on their nutritional genomics in maize. A bioinformatics approach, however, has been reported by Chauhan [205] toward the identification of candidate genes for zinc and iron transporters in maize using the sequence data available for maize and the iron and zinc metabolism information derived from other plant species such as Arabidopsis.

Conventional biofortification breeding efforts are attempting to enhance iron and zinc content by using maize genetic resources. At CIMMYT, we have analyzed more than 1000 improved maize genotypes and 400 landraces from different environments. However, we found little variation in grain for iron levels (average ppm) [206,207] and only moderate variation for zinc concentration in grain (15–35 ppm).

PROTEIN

The major maize seed storage proteins, zeins, are deficient in two essential amino acids, lysine and tryptophan, and therefore contribute to the poor nutritional quality of maize. Kernels with reduced levels of zein proteins (through mutation or genetic engineering) have been shown to have increased levels of lysine and tryptophan [208]. The presence of a naturally-occurring mutant gene, the opaque-2 (o2), results in increased concentrations of lysine and tryptophan in maize grain, which has been named quality protein maize (QPM). This equates to a biological nutritional value for QPM protein that is 90%

equivalent to that of protein in milk, whereas that of regular maize grain in only about 40% [209].

Gene expression studies, QTL analysis, and progress in proteomics, transcriptomics and conventional breeding have helped to elucidate the QPM trait [208, 210]. Breeding of QPM cultivars requires manipulation of three genetic systems: (1) the opaque-2 (o2) gene must be in its homozygous recessive form, thereby reducing the rate of transcription of genes encoding zein proteins, which contain very small quantities of lysine and tryptophan; (2) modifier genes of the o2 gene must be selected to modify the undesirable soft and chalky (opaque) kernel features that are typical of opaque-2 maize; (3) additional (non-o2) genes affecting lysine and tryptophan concentration in grain must be selected to ensure that concentrations of these amino acids are within the high range of variation observed for maize [211, 212].

By genetic engineering, a dominant opaque phenotype has been obtained by reduction of the zein proteins in the grain showing increased lysine and tryptophan contents [213, 214]. Most recently, Houmard et al. [215] have reported the increase of lysine in maize grains by endosperm-specific suppression of lysine catabolism using RNA interference (RNAi). A different approach to increase protein quality in maize using genetic engineering has been the introduction of the gene encoding amarantin. Amarantin is a protein from the Amaranth plant, which is known to be well balanced in its amino acid content. An increase of 8% to 44% in essential amino acids was observed for maize transformed with the amarantin encoding gene [216]. However, the efficacy of this maize for target population is still to be demonstrated.

Carotenoids

Given the general nutritional interest in carotenoids for intermediate and end products of maize, the carotenoid metabolic pathway has been intensively researched. All yellow maize contains carotenoids, although the fraction of carotenoids with provitamin A activity (-cryptoxanthin, - and -carotene, which can be converted to vitamin A) is typically small (15% to 18% of the total carotenoid fraction) compared to zeaxanthin and lutein (around 45% and 35%, respectively) [217, 218]. Due to the general vitamin A deficiency in developing countries, efforts have been concentrated on increasing the amount of provitamin A carotenoids in staple crops including maize [219]. There is considerable scope for breeding maize with enhanced provitamin A concentrations by shifting carotenoid biosynthesis to favour provitamin A versus other types of carotenoids, and hence enhancing nutritional value [218, 220, 221].

Analyses of genotypes with yellow to dark orange kernels have identified large variation in the number of provitamin A molecules [220, 222] and their carotenoid profiles. At CIMMYT, we have analyzed carotenoid profiles for more than 1000 tropical genotypes and identified promising materials with provitamin A concentrations (ca. 8 g/g) and/or carotenoid profiles that could be used in breeding progams to increase total provitamin A content in maize grain. To date we have observed no consistent trend in the origin of maize genotypes with the highest provitamin A concentrations; the best materials include pale-yellow temperate, dark-yellow highland tropical, and intense-orange tropical lines.

There are many carotenoid phenotypic mutants in maize that have been associated with cloned genes. Psy1(phytoene synthase) was cloned by transposon tagging and mapped to chromosome 6 [223], and in the presence of the resulting Yellow 1 (Y1) gene product, carotenoids are produced in the endosperm tissue, yielding the yellow endosperm phenotype. The other allelic form and psy2 are not expressed in endosperm [72, 223, 224]. In addition, QTL analyses have shown that candidate genes Psy1 and Zetacarotene desaturase (Zds) are associated with variation of individual and total carotenoid contents [225].

Using an RIL population derived from a cross between By804 and B73, 31 QTL including 23 for individual and eight for total carotenoids were detected [226]. Much of the phenotypic variation in carotenoids could be explained by two loci, and the QTL for carotenoids elucidated the interrelationship among these compounds at the molecular level. A gene targeted marker in the candidate gene psy1 tightly linked to a major QTL was identified explaining 6.6%–27.2% of phenotypic variation for levels of carotenoids. Fraser and Bramley [227] confirmed that LycE controlled the ratio of carotenoids in zeaxanthin (with provitamin A intermediates) and lutein branches, being a key enzyme in the provitamin A content of maize. Through association analysis, linkage mapping, expression analysis, and mutagensis, Harjes et al. [29] showed that variation at the lycopene epsilon cyclase (lcyE) locus alters flux down -carotene versus -carotene branches of the carotenoid pathway. Four natural lcyE polymorphisms explained 58% of the variation in these branches and a threefold difference in provitamin A compounds. Another gene in the pathway, carotene hydroxylase enzyme (CrtR-B1), has also been cloned and analyzed (Jianbing Yan, unpublished). A recent report in genetic transformation, as discussed in Section 6, has used particle bombardment to generate multivitamin maize with significantly increased contents for -carotene, ascorbate, and folate [178]. This is a very important proof of concept for genetic manipulation of distinct metabolic pathways. However, appropriate and fast strategies have to be developed now to obtain the useable products and the desired impact in the target countries.

GENOMICS-ASSISTED BREEDING

While almost all maize plants grown in the fields of developed countries are hybrids, various types of maize cultivars are grown in developing countries, including hybrids, synthetic cultivars, open pollinated varieties (OPVs), and landraces. In addition to the target type of maize cultivar, breeding objectives are dominated by the target environment. High yield potential and good grain quality are primary selection criteria for maize breeding programs across both tropical and temperate areas. However, tropical maize breeders are required to address some very specific abiotic and biotic stress tolerance objectives that are rarely important for temperate maize breeding programs.

The multinational seed companies are now routinely using applied genomic tools to (i) dissect the genetic structure of relevant maize germplasm to understand gene pools and germplasm (heterotic) groups, (ii) provide insights into allelic content of genetic resources for potential use in breeding, (iii) screen early generation breeding populations to select segregants with desired combinations of marker alleles associated with beneficial traits (in order to reduce the scale of costly phenotypic evaluations), (iv) pyramid favorable genes/alleles from different germplasm sources through marker-assisted recurrent selection in order to improve genetic gains, and (v) establish genetic identity (fingerprinting) of their products [61, 228–235]. MAS has been successfully applied by private sector maize cultivar development aimed at recovery of an ideal genotype defined as a mosaic of favorable chromosomal segments. Parental genotypes are selected to provide specific components of this mosaic including favorable alleles for multiple complex traits such as yield, biotic and abiotic stress resistance, and quality attributes [234, 236–238]. Using these approaches, the commercial breeding programs report a doubling in the rate of genetic gain over phenotypic selection (PS) in maize [233, 237].

A large proportion of the successful reports of MAS in maize have involved MABC or advanced backcross QTL (AB-QTL) for introgression of favorable alleles (foreground selection) and for accelerating the recovery of the recipient genotype in the remainder of genome (background selection, see Table 1). In theory, the relatively easy target traits for MAS include grain quality genes and major disease resistance genes. However, there are also some reports regarding complex traits. A good example was the improvement of the B73 Mo17 hybrid through marker-assisted enhancement with elite alleles from Tx303 and Oh43 [239]. On the basis of four years of testing, yields of the best "enhanced" B73 enhanced Mo17 hybrids exceeded the original hybrid and high yielding commercial hybrids by 8% to 10%. In another case, the drought susceptible line CML247 was improved with five favorable alleles from the drought tolerant line Ac7634 through four generations of MABC. As a result, the best five MABC-derived hybrids yielded, on average, at least 50% more than the control hybrids under water stress conditions [240, 241].

Table 1: Examples of MAS applications in maize improvement

Traits	**Molecular markers**	**Note**	**References**
Grain yield, ear number, plant height	RFLP	Increase of hybrid yield by enhancing both parents with introgression of favorable alleles	[239, 254]
Grain yielding, percent stalk lodging	RFLP	Early generation test of inbreeding process	[255]
Grain yield, earliness	RFLP	Foreground selection for three loci of earliness and grain yield and background selection	[256]
Grain yield	SSR	Four yield QTL were selected as target introgressions in the developmentof BC3 families	[257]

Seeding emergency	RFLP	Seedling emergence of three elite commercial sweet corn inbreds has been improved by introgression of three alleles with marker-assisted backcross	[258]
Drought tolerance	RFLP, SSR	Five favorable alleles related to drought tolerance were successfully introgressed into the recurrent parent	[240]
Fertility restorer gene	SSR	Selection of Rf3 gene	[259]
Southwestern corn borer	RFLP	A total of five loci traced during backcross	[260]
Stalk strength, second-generation European corn borer	RFLP	Comparison of effectiveness between phenotypic selection and MAS on stalk strength and second-generation European corn borer resistance	[261]
Quality protein maize (QPM)	SSR	Both foreground selection for the o2 gene and background selection for restoration of recurrent genome	[262]
Quality protein maize (QPM)	SSR	o2 gene was introgressed into herbicide resistant elite maize inbred lines using three SSR markers	[263]
Drought tolerance	RFLP	Inbred line CML247 was improved for drought tolerance through four cycles of MABC	[241]

Despite all the recent technological breakthroughs, there are limited reports on the overall contribution of genomics-assisted breeding of quantitative traits in released cultivars, particularly in public sector breeding programs [61, 242]. Progress in rice has been better documented in literature [243]. For example, a gene underlying a large-effect QTL contributing to submergence tolerance in rice has been cloned and introgressed into a wide range of cultivars [244]. Although there is very limited specific information on the successes of molecular breeding, the first commercial products of molecular breeding (rather than limited MAS) are already being released to the market by several major multinational breeding companies. The first hybrid maize cultivars developed through molecular breeding by Monsanto were released for commercial production in the USA for the 2006 cropping season, and it is estimated that, by 2010, over 12% of the commercial crop in the USA will be derived from products of molecular breeding [245].

Yield and Heterosis

Dominance, overdominance, and epistasis have all been proposed to have a role in the genetic basis of superior hybrid performance. The dominance model attributes increased vigor to the interaction of favorable dominant alleles from both parents when combined in the hybrid [246], whereas the over dominance model postulates the existence of loci at which the heterozygous state is superior to either homozygote [52]. Evidence for the role of epistasis (interaction of the favorable alleles at different loci contributed by the two parents) in hybrid vigor has also been reported [247–249]. The genetic basis of heterosis, heterotic groups, hybrid prediction and hybrid performance, relationships between heterozygosity and genetic distance with hybrid performance and heterosis, and use of MAS in hybrid breeding has been reviewed elsewhere [229].

Harnessing heterosis has been a corner stone of maize breeding for nearly a century and has been more extensively used than for any other crop. Therefore, development of a reliable method for predicting hybrid maize performance without generating and testing hundreds or thousands of test-cross combinations has been the goal of numerous studies, using both marker data and a combination of marker and phenotypic data [239]. Most of these studies concentrated on probing the general correlation between genetic distances revealed by molecular markers and hybrid performance. However, without identifying the association of markers with specific genes for hybrid vigor and performance, the correlation revealed by random markers has only been able to provide very limited predictive power.

The definition of heterotic groups and heterotic patterns in temperate hybrid maize breeding has contributed to large increases in yield [250]. Reciprocal recurrent selection breeding programs have proven effective in the improvement of heterotic groups for a systematic exploitation of heterosis, as they maximize selection gains within a heterotic group while at the same time maximizing differences between heterotic groups. Clear characterization of genetic diversity among maize inbred lines derived from different origins can increase the efficiency of predicting good hybrid combinations for developing new inbred lines. In temperate maize such as the US Corn Belt germplasm, a clear heterotic pattern (Reid Stiff Stalk versus Lancaster) was established early on and inbred lines such as B73 and Mo17 from the two heterotic groups were chosen as testers for the selection of new maize inbred lines. Today two heterotic groups (Stiff-Stalk (SS) versus Non-Stiff-Stalk (NSS)) are clearly distinguished. For tropical germplasm, there are two apparent heterotic groups: SS-Tuxpeno versus NSS-Non-Tuxpeno.

Following successful deployment of hundreds of OPVs in the 1970s and early 1980s, the CIMMYT maize program began the development of hybrid maize to meet the needs of an increased number of developing country farmers who were eager to switch to this type of variety. In the 1990s, 10 pairs of heterotically distinct genotypes were identified or generating in subtropical, midaltitude, and highland populations to provide heterotic pools for breeders targeting each of these environments. More recently, testers from each population have been used to identify the hybrid performance of inbreds from the partner populations and help assign new inbred lines to appropriate heterotic groups.

Molecular markers are useful for helping to define heterotic groups and to examine the relationships among inbred lines at the DNA level. Various types of molecular marker have been used to investigate relationships among inbred maize lines from different heterotic groups, including tropical maize [13]. Markers are also useful to assign new lines to new or currently existing heterotic groups.

Abiotic Stresses

New molecular marker technologies and comprehensive gene expression profiling methods provide opportunities to direct the breeding of improved genotypes that provide stable grain yield under varied suboptimum environmental conditions [251]. In this context physiological genomics may improve our ability to manipulate the genome of crop plants in order to improve their adaptation to stresses [252, 253].

QTL analysis has improved our knowledge of the genetic basis of a number of morphophysiological traits involved in the response to different abiotic stresses in maize. A number of candidate genes mapped near to QTL regulating important morphophysiological traits and grain yield have already been identified in maize. An updated compilation of mapped QTL and major genes associated with abiotic stress tolerance in maize and other plants is available at http://www.plantstress.com/biotech/index.asp?Flag=1. Among the most important traits are yield, flowering and phenological parameters, plant height, ear number, photosynthesis, chlorophyll fluorescence and leaf ABA concentration, and root traits such as adventitious root formation for waterlogging and chilling tolerance as well as root hair traits for phosphorus deficiency and aluminum tolerance. Khavkin and Coe [264] hypothesized that many apparent QTL of major effect in maize are in fact clusters of genes (e.g., homeotic genes and other genes encoding for transcription factors) regulating development and that many plant reactions to abiotic stresses rely on such gene clusters.

Drought tolerance has been the most difficult abiotic stress for breeders to make rapid, substantial, and consistent advances. The genetic basis of the molecular, cellular, and developmental responses to drought involves many gene functions regulated by water availability [242]. In that context, genomics-based approaches could help access to agronomically desirable alleles present at QTL which eventually may help to improve the drought tolerance and yield of crops under drought. The identification and cloning of genes at target QTL may further broaden our understanding of the genetic, physiological, and functional bases of drought tolerance [265].

A major effect of water stress in maize is a delay in silking, resulting in an increase in the anthesis-silking interval (ASI), which leads to substantial yield reduction or even complete crop failure. Thus, developing maize lines with a short ASI has been an important goal for drought tolerance breeders. QTL influencing flowering time and ASI have been identified [266–268] and the availability of molecular markers linked to five QTL for ASI enabled lines with a reduced ASI to be selected [55, 240]; however, the yield advantage was only evident under very severe stress conditions. In addition, QTL for silking date, grain yield and yield stability under field conditions have been reported [269]. Since vegetative plant growth is strongly affected by drought stress, emphasis has also been placed on identification of QTL for the response of leaf elongation rate to soil moisture, temperature and evaporative demand [270]. Several factors appear to have confounded the detection of QTL in maize that could be useful in marker-assisted development of drought-tolerant cultivars. Among them has been a tendency to examine crosses between lines that were not agronomically elite and/or did not exhibit extreme differences in yield under stress. Furthermore, many studies have used relatively small mapping populations and have achieved low precision in phenotyping, factors that are well known to result in low power of QTL detection and severe bias in the estimation of QTL effects [271]. Other studies have relied on an inadequate phenotyping in terms of how, when and what traits are measured, thereby limiting their impact [272]. Thus, the use of MAS for improving complex traits remains a challenge for crop breeders [273], at least in the public sector [61]. While the genetic dissection of crop performance in drought-prone environments has greatly benefited from the use of DNA markers [241, 242, 274, 275], the outputs have not been routinely implementation in practical breeding programs. QTL are often germplasm-specific and the costs for applying MAS for many QTL of small effect may be greater than those of conventional cross-breeding. The challenge for molecular breeders is to identify QTL of major effect that are independent of genetic background and to devise more effective breeding approaches for the application of the resultant markers, such as pedigree selection. This approach has shown promise in rice [276, 277] as well as in durum wheat [275]. In both cases, results suggest that breeding for traits affecting yield potential can be translated into better performance under drought stress [265].

MAS for drought-related traits based on genetic mapping information should preferably target "major" QTL with a sizeable effect, consistent across germplasm and with a limited interaction with the profile of water availability. In maize, however, QTL studies in the past have not identified any QTL with sufficiently large effects to be effectively used in MAS programs. Marker-assisted recurrent selection provides an alternative approach to improve drought tolerance through genome-wide selection or selection based on an index constructed from a set of markers associated with different traits or trait components. This approach has been proven to be successful in private sector breeding programs [234, 235] and is being extensively implemented in the Drought Tolerance Maize for Africa project using chip-based SNP markers (http://dtma.cimmyt.org).

Quality

The molecular breeding of QPM has been reviewed elsewhere [278]. There are three markers available for MAS in QPM breeding (phi057, phi112, and umc 1066). One of the markers (phi112) provides only dominant information, and thus can only be used to identify genotypes that do not contain a recessive o2 allele. The other two markers exhibit codominant polymorphism between normal and QPM inbreds [262]. Even with MAS, due to the polygenic nature of QPM and the current status of available markers, protein content and quality must still be monitored using biochemical methods. The most difficult aspect of QPM molecular breeding is the selection for minor loci controlling modification of lysine and tryptophan levels in o2o2 backgrounds, of which over eight loci have been mapped in various studies [212]. At least two and likely more loci affect the modification of the endosperm hardness in o2o2 backgrounds [279]. Several major o2 modifier QTL have been mapped to chromosomes 1, 7, and 9. A microarray hybridization performed with RNA obtained from true breeding o2progeny with vitreous and opaque kernel phenotypes identified a small group of differentially expressed genes, some of which mapped at or near the o2 modifier QTL. Several of the genes were associated with ethylene and ABA signaling which suggests a potential linkage of o2 endosperm modification with programmed cell death [280].

Using SSR-based MABC breeding, maize lines have been developed that have twice the amount of lysine and tryptophan than the native lines and up to 95% of the recurrent parent genome [262]. In African maize breeding,o2 allele specific SSR markers were used to convert herbicide resistance maize lines into QPM which is the equivalent of modified o2 phenotype. Using the three SSR markers, the result showed that 97% of the lines wereo2. Conventional methods using a light table or light box and MAS obtained comparable results [263]. However, application of SSR markers and the FTA DNA extraction technology offered the breeder a fast, reliable and less labor intensive method of screening QPM maize during the early growing stages instead of having to wait to screen the kernels on the light table after harvesting. Unfortunately, o2 selection alone is not sufficient as breeders must also select for the tryptophan and lysine modifiers using biochemical analysis.

Using a new high lysine mutant, o16, which contains similar levels of lysine to o2 mutants [281], MAS for combining favorable alleles of both o2 and o16 would facilitate development of new high lysine maize cultivars once the genetic effect ofo16 is confirmed under different genetic backgrounds [278].

Plant oils have been receiving increasing attention as an important renewable resource for biodiesel production and for dietary consumption by humans and livestock. A high-oil QTL (qHO6) affecting maize seed oil and oleic-acid content has been shown to encode an acyl-CoA : diacylglycerol acyltransferase (DGAT1-2), which catalyzes the final step of oil synthesis. A phenylalanine insertion in DGAT1-2 at position 469 (F469) was responsible for the increased oil and oleic-acid content. Ectopic expression of the high-oil DGAT1-2 allele increased oil and oleic-acid content by up to 41% and 107%, respectively [282]. This work provides insights into the molecular basis of natural variation of oil and oleic-acid contents in plants and highlights DGAT as a promising target for increasing oil and oleic-acid content in other crops.

MAS and phenotypic selection (PS) were applied to three $F_{2:3}$ populations as base (starting) populations (C_n) with either the su1, se1, or sh2 endosperm mutations [283], which are related to sweetcorn quality. Of the 52 paired comparisons made between composite populations derived through MAS or PS, MAS resulted in significantly higher gains than PS in 38% of comparisons across the three C_1 composite populations, while PS was significantly greater in only 4% of the comparisons. The average MAS and PS gain across all composite populations and selected traits, calculated as a percentage increase from the randomly selected controls, was 10.9% and 6.1%, respectively. Another example is the development of an MAS system for two genes in the carotenoid pathway that have been cloned (as discussed in the previous section). Inexpensive PCR-based markers for provitamin A are being developed and validated at CIMMYT from the cloned gene sequence (unpublished results, CIMMYT) using seed DNA-based genotyping method [284], which will enable developing-country breeders to more effectively produce maize cultivars with higher provitamin A levels. A molecular breeding platform based on MAS using SNP-chips has been developed and used for linkage mapping and LD-based analysis and molecular breeding for both qualitative and quantitative traits [86, 285].

Genomic Databases and Informatics Tools

In the past few years, several important databases have been built focused on data from maize genomics research (Table 2). One of the most important collection of genomic databases and informatics tools is MaizeGDB (http://www.maizegdb.org). Among the data sets included in MaizeGDB are sequences, including integration with various contig assemblies; publication references; detailed genetic, physical, and cytogenetic maps; QTL mapping results; mutants; genes; primers; and a wealth of other data types. MaizeGDB includes integrated tools for map comparisons, sequence similarity searches, and comparisons with and links to other databases, such as Gramene and NCBI. MaizeGDB provides web-based curation tools that enable researchers to edit and annotate their own data and to enter new data into MaizeGDB directly. MaizeGDB also provides informatics support for maize community initiatives such as the annual Maize Genetics Conference and community-wide workshops, and maintains data for maize community research projects. To have easy access to systematized information about all known QTL for traits of interest, coupled to other information about germplasm, nearby loci and sequence information, MaizeGDB began curating QTL information from literature in the mid 90's. To permit this work to continue at MaizeGDB, a new Web accessible curation interface has been designed and implemented. The new design accommodates a legacy trait hierarchy developed at MaizeGDB and recently harmonized with the rice Trait Ontology at Gramene, and trait descriptors used by GRIN (the Germplasm Resources Information Network).

Table 2: Databases and resources related to maize

Name and link	Description
MaizeGDB http://www.maizegdb.org/	A community database for biological information about the crop plant Zea mays ssp. mays. Genetic, genomic, sequence, gene product, functional characterization, literature reference, and person/organization contact information are among the data types accessible (see Section 9 for further details)
MAGI http://magi.plantgenomics.iastate.edu/	Maize assembled genomic island (MAGI) reports the results of a maize genome assembly project being conducted by the Aluru, Ashlock and Schnable research groups

Gramene http://www.gramene.org/	A resource for comparative grass genomics including: genomes, protein, maps, markers, traits, genetic diversity, biochemical pathway, literature, and so forth
Maizesequencehttp://www.maizesequence.org/index.html	Provides sequence and annotation of the Zea mays ssp. Mays genome resulting from the maize genome sequencing project
mtmDB http://mtm.cshl.edu/	A transposon library in maize for the purpose of targeted selection of gene knockouts
Maize tilling http://genome.purdue.edu/maizetilling/	A website for maize tilling project that collects and distributes maize tilling mutant and seeds
NCBI Maizehttp://www.ncbi.nlm.nih.gov/mapview/map_search.cgi?taxid=4577	Provides all of the accessions from Zea mays including: genomics, cDNA, EST, GSS, HTG sequences
SAM http://sam.truman.edu/	A database for maize shoot apical meristem expression
Helitonshttp://genomecluster.secs.oakland.edu/helitrons/	A catalog of discovered helitrons
Grassius http://www.grassius.org/	Provides a collection of databases for transcription factors, promoters and cis-regulatory elements across the grasses, including maize, sugarcane, sorghum, and rice
Panzea http://www.panzea.org/	A database for molecular and functional diversity of the maize genome
MaizePLEX http://www.plexdb.org/plex.php?database=Corn	A miame-compliant and plant ontology enhanced expression database for maize microarray data
TIGR Maize http://maize.tigr.org/	Provides links to the NSF-funded consortium for maize genomics project and includes sequence, assembly, and annotation data, and links to the maize gene index
MaizecDNA http://www.maizecdna.org/	Released 30000 FLcDNA from diverse B73 tissues samples
ZmGDB http://www.plantgdb.org/ZmGDB/	Provides a convenient sequence-centered genome view for Zea mays, with a narrow focus on gene structure annotation
PML http://chloroplast.uoregon.edu/	A genetic resource in maize, that is, tailored for studies of chloroplast biogenesis
VPhenoDBShttp://medbio.cecs.missouri.edu/VPhenoDBS/	Web-based visual phenotypic information management system to allow simultaneously query phenotype data by image example, sequence, ontology, genetic and physical map information, and text annotation

The Maize Assembled Genomic Island site (MAGI, http://magi.plantgenomics.iastate.edu) is a resource for maize genome assembly, annotation and mapping, which assembled a large number of maize genomic sequences primarily composed of gene-enriched GSSs (genomic survey sequences), random Whole Genome Shotgun (WGS) sequences, and BAC shotgun reads [286]. GBrowse, a component of GMOD (a Generic Model Organism Database Toolkit), is used to display annotated assemblies. The MAGI website can serve as a community resource for map-based cloning projects as well as for analyses of genome structure and comparative genomics. The FPC resources at the Arizona Genomics Institute provide information on the status of the agarose FPC map. WebFPC and its associated tools provide online access to the contigs. Much effort on maize genome sequence annotation has been made at http://www.maizesequence.org, linking maize sequence to physical and genetic maps, and providing computational annotation of predicted genes—and this portal will see widespread use as the sequence is completed and groups start to focus on functional analysis.

To enable biologists to simultaneously query phenotype data by image example, sequence, ontology, genetic and physical map information, and text annotations, a web-based visual phenotypic information management system, VPhenoDBS (medbio.cecs.missouri.edu/VPhenoDBS), is being developed [287]. The database framework consists of five modules: a system to extract and quantify low level features from phenotypic images, a high-dimensional database indexing system to manage and cluster images for real-time retrievals, a linking hub to correlate visual features already attributed to a given locus with relevant genetic and physical maps, a text mining and ontology utilization system for parsing annotations, and a results visualization system. This system may be integrated with a fully automatic high-throughput screening system, the Scanalyzer 3D, as presented for complete plants like maize, rice, Arabidopsis, poplar tree, barley or wheat in the greenhouse, combining information from all 3 dimensions (http://www.lemnatec.de/scanalyzer_gh.htm). This screening system is able to 3-dimensionally screen up to 4000 plants per day efficiently and precisely. With the Scanalyzer 3D a wide range of visual evaluation parameters of plants can be sampled for a complete, reproducible and nondestructive analysis free of subjective influences.

There are many software and decision support tools developed in plant genomics including those for germplasm evaluation, breeding population management, genetic map construction, marker-trait association analysis, MAS, GEI analysis, breeding design and simulation, and information management [288]. Table 3 lists some tools, which include QTL meta-analysis, QTL and gene comparative analysis, population structure and kinship evaluation, and association mapping that are particularly valuable in maize genomics research and molecular breeding.

Table 3: Bioinformatics tools for maize genomic and functional research

Name and link	Description
MetaQTL http://bioinformatics.org/mqtl/	A java package designed to perform the integration of data from the field for gene mapping experiments
BioMercator http://cms.moulon.inra.fr/index.php?option=com_content&task=view&id=13&Itemid=43	Genetic maps and QTL integration
CMTV http://www.ncgr.org/cmtv/	An integrated bioinformatics tool to construct consensus maps and compare QTL and functional genomics data across genomes and experiments
QTL-Finder http://gqtl.maizecenter.cn	A bioinformatics tool for QTL integration, comparison and candidate gene discovery across genomes and experiments
TWINSCAN http://mblab.wustl.edu/query.html	Improved gene prediction performance for maize and rice
GDPChttp://www.maizegenetics.net/gdpc/index.html	The genomic diversity and phenotype connection (GDPC) simplifies access to genomic diversity and phenotype data, thereby encouraging reuse of this data. GDPC accomplishes this by retrieving data from one or more data sources and by allowing researchers to analyze integrated data in a standard format. GDPC provides access to genomic diversity data such as SNPs, SSRs, and sequences, and phenotypic data that may be collected in field, genetic, or physiological experiments
TASSEL http://www.maizegenetics.net/index.php?page=bioinformatics/tassel/index.html	A software package to evaluate trait associations, evolutionary patterns, and linkage disequilibrium
PowerMarker http://statgen.ncsu.edu/powermarker/	A comprehensive set of statistical methods for genetic marker data analysis, designed especially for SSR/SNP data

RepMinerhttp://jestill.myweb.uga.edu/RepMiner.htm	Takes a graph theory approach to the identification and assembly of transposable elements from small DNA fragments resulting from subcloning bacterial artificial chromosome libraries
SPAGeDihttp://www.ulb.ac.be/sciences/ecoevol/spagedi.html	Spatial Pattern Analysis of Genetic Diversity (SPAGeDi) is a new computer package—replacing AutocorG that was distributed to a limited extent—primarily designed to characterize the spatial genetic structure of mapped individuals and/or mapped populations using genotype data of any ploidy level
Structurehttp://pritch.bsd.uchicago.edu/structure.html	A free software package for using multilocus genotype data to investigate population structure
TEnest http://bak.public.iastate.edu/te_nest.html	Automated chronological annotation and visualization of maize nested transposable elements

Two software packages that are more specific to maize will be discussed here. The first is TEnest, which was developed to facilitate identifying repetitive sequences and reconstructing separated sections to provide full-length repeats and, for long-terminal repeat (LTR) retrotransposons, calculating age since insertion based on LTR divergence [289]. Considering that 85% of the maize genome consists of transposable elements (TEs), with more than 70% of TEs found nested within one another, an accurate nested TE identification tool for complete annotation of the maize genome is needed. TEnest contains an up-to-date database of maize canonical TEs and their associated LTRs, if applicable. The second software we would like to highlight is integrated software for SNP discovery in maize [290]. The development of software tools to aid researchers in the SNP discovery process across several maize, teosinte, and Tripsacum lines has been the focus. An integrated set of tools consisting of a relational database and applications for data loading, editing and reporting has been developed (http://www.panzea.org). All stages of SNP discovery from tracking sequences, alignment generation, alignment editing, and reporting are covered. Central to this system is an intuitive, quality score-based alignment editing tool designed to simplify manual editing of the highly polymorphic and complex Zea alignments.

OUTLOOK

Genome-wide scans for genetic mapping and whole genome sequence-assisted marker development and application have now become possible in maize. The process will be accelerated by high-throughput technologies for both phenotyping and genotyping, and will be facilitated by bioinformatics and decision support tools [61]. We anticipate that large-scale genomics-assisted marker development and gene discovery will be routinely applied in both private and public sectors for traits of economic importance, especially where individual genes or QTL of large effects can be identified that are not significantly influenced by environmental factors. Conversely, to make improvements in traits controlled by many genes, each having minor effects and all having large GEI, will require a much more complicated approach. In particular, well-adjusted strategies and strengthened multidisciplinary collaboration will be needed, and in many cases the solutions will be trait-specific in development and population-specific in application.

Considering the progress in the various "omics" areas and the integration of different disciplinary applications facilitated by bioinformatics, as well as high-throughput genotyping approaches combined with automation, MAS will gradually evolve into more holistic "genomics-assisted" breeding strategies. Since genomics resources in maize are among the best of the major crop species, the role of genomics is set to become more and more important in maize breeding. However, conventional selection will remain a vital element of the process to finally confirm the best candidate genotypes for progression into the advanced stages of crop improvement and cultivar selection. Thus, it will not be a case of MAS replacing parts of traditional breeding programs, but rather an increasing reliance on genomics data alongside other technology interventions in an ever evolving and refining breeding system. Genomics-assisted breeding systems will be evaluated in terms of their ability to increase the scope of breeding goals, to provide new added-value traits, to decrease the cost of breeding programs, and to improve the pace of developing new cultivars, and finally to enhance impact of resultant products to command increasing areas of production.

Transcriptomics is a field that may experience significant innovations in the near future with increasing impact on physiological studies of maize. Monitoring large-scale changes in transcript profiles may eventually allow for the identification of transcript networks accounting for GEIs in different genotypes [291]. However, there are several major limitations of microarray-based expressional studies, for example: (1) low-abundance mRNA may not be represented by most arrays nor detected upon hybridization; (2) the correlation between the level of mRNA and the products of their translation or their biological effect can be low. Genome-wide transcript profiling may encounter a similar history as genome mapping that started with great impacts for the traits controlled by major genes or QTL without significant GEI, but still faces challenges for manipulating complex traits.

Maize as a multipurpose crop of importance in all global crop production regions will continue to play a leading role in shaping the future of crop improvement and production systems. Advances in maize genomics, breeding and production will have significant impact on the lives of a large proportion of the world's population [292]. Balancing consumer demand for various usages of maize with different maize production practices will be critical for maintaining both sustainability of cropping systems, food security, feed and fodder supply and bioenergy demands. It is important to note that the edible portions of food crops are not the most desirable plant portion for creating biofuel in many plants including maize. The current processes of generating biofuel from maize must be improved either by more efficient use of the maize grain or better conversion of cellulose contained in the stalks into biofuels. The latter, so-called second generation biofuels, have not yet been fully developed for any crop but hold much promise. Of particular importance is that this approach would not directly compete with much-needed food supply. Although it would compete with sustainable conservation agriculture production systems and the increasing demand for fodder to support the rapidly expanding livestock industry in developing countries. There is clearly no quick and easy fix to this dynamic situation and much careful thought and action will be required to properly manage the situation [293].

As maize is a fundamentally important commodity in both developed and developing countries and used in many varied ways, North-South collaborations in maize genomics should be strengthened for scientists in both theoretical and applied genomics fields. We can safely assume that the rapid developments in maize genetics and genomics, although currently based mainly on temperate maize germplasm, will be transferable and increasingly valuable for improvement of tropical and subtropical maize which is the important crop for food security in developing countries. With the resolution of many practical, logistical and genetic bottlenecks in MAS [61], including development of seed DNA-based genotyping [284], and the ongoing development of powerful decision support tools, it can be expected that genomics-assisted approaches will increasingly become a routine component of breeding programs of private and public sectors worldwide.

ACKNOWLEDGMENTS

The authors would like to thank Drs. Rodomiro Ortiz, Kevin Pixley, and Gary Atlin for their critical review of the manuscript. Genomics and molecular breeding research at CIMMYT has been supported by the Rockefeller Foundation, Bill and Melinda Gates Foundation, European Community, Generation Challenge Program and HarvestPlus Challenge Program, and through other attributed or unrestricted funds provided by the members of the CGIAR and national governments of USA, Japan, and UK. Shibin Gao was supported by the Scientific Research Foundation for the Returned Overseas Chinese Scholars, State Education Ministry, China.

REFERENCES

1. R. L. Paliwal, G. Granados, H. R. Lafitte, A. D. Violic, and J. P. Marathee, Tropical Maize: Improvement and Production, Food and Agriculture Organization of the Unites Nations, Rome, Italy, 2000.
2. J. Messing, A. K. Bharti, W. M. Karlowski, et al., "Sequence composition and genome organization of maize," Proceedings

of the National Academy of Sciences of the United States of America, vol. 101, no. 40, pp. 14349–14354, 2004.

3. G. Haberer, S. Young, A. K. Bharti, et al., "Structure and architecture of the maize genome," Plant Physiology, vol. 139, no. 4, pp. 1612–1624, 2005.
4. E. S. Buckler, B. S. Gaut, and M. D. McMullen, "Molecular and functional diversity of maize," Current Opinion in Plant Biology, vol. 9, no. 2, pp. 172–176, 2006.
5. J. C. Reif, X. C. Xia, A. E. Melchinger, et al., "Genetic diversity determined within and among CIMMYT maize populations of tropical, subtropical, and temperate germplasm by SSR markers," Crop Science, vol. 44, no. 1, pp. 326–334, 2004.
6. P. Dubreuil and A. Charcosset, "Genetic diversity within and among maize populations: a comparison between isozyme and nuclear RLFP loci," Theoretical and Applied Genetics, vol. 96, no. 5, pp. 577–587, 1998.
7. M. L. Warburton, X. Xianchun, J. Crossa, et al., "Genetic characterization of CIMMYT inbred maize lines and open pollinated populations using large scale fingerprinting methods," Crop Science, vol. 42, no. 6, pp. 1832–1840, 2002.
8. Y. Vigouroux, J. C. Glaubitz, Y. Matsuoka, M. M. Goodman, J. Sánchez G., and J. Doebley, "Population structure and genetic diversity of new world maize races assessed by DNA microsatellites," American Journal of Botany, vol. 95, no. 10, pp. 1240–1253, 2008.
9. M. L. Warburton, J. C. Reif, M. Frisch, et al., "Genetic diversity in CIMMYT nontemperate maize germplasm: landraces, open pollinated varieties, and inbred lines," Crop Science, vol. 48, no. 2, pp. 617–624, 2008.
10. H. Lu and R. Bernardo, "Molecular marker diversity among current and historical maize inbreds,"Theoretical and Applied Genetics, vol. 103, no. 4, pp. 613–617, 2001.
11. F. J. Betrán, J. M. Ribaut, D. Beck, and D. Gonzalez de León, "Genetic diversity, specific combining ability, and heterosis in tropical maize under stress and nonstress environments," Crop Science, vol. 43, no. 3, pp. 797–806, 2003.

12. K. Liu, M. Goodman, S. Muse, J. S. Smith, E. Buckler, and J. Doebley, "Genetic structure and diversity among maize inbred lines as inferred from DNA microsatellites," Genetics, vol. 165, no. 4, pp. 2117–2128, 2003.
13. X. C. Xia, J. C. Reif, D. A. Hoisington, A. E. Melchinger, M. Frisch, and M. L. Warburton, "Genetic diversity among CIMMYT maize inbred lines investigated with SSR markers: I. Lowland tropical maize," Crop Science, vol. 44, no. 6, pp. 2230–2237, 2004.
14. X. C. Xia, J. C. Reif, A. E. Melchinger, et al., "Genetic diversity among CIMMYT maize inbred lines investigated with SSR markers: II. Subtropical, tropical midaltitude, and highland maize inbred lines and their relationships with elite U.S. and European maize," Crop Science, vol. 45, no. 6, pp. 2573–2582, 2005.
15. A. E. Melchinger, "Genetic diversity and heterosis," in The Genetics and Exploitation of Heterosis in Crops, J. G. Coors and S. Pandey, Eds., chapter 10, CSSA, Madison, Wiss, USA, 1999.
16. D. N. Duvick, J. S. C. Smith, and M. Cooper, "Changes in performance, parentage, and genetic diversity of successful corn hybrids, 1930–2000," in Corn: Origin, History, Technology, and Production, C. W. Smith, J. Betran, and E. C. A. Runge, Eds., pp. 65–97, John Wiley & Sons, New York, NY, USA, 2004.
17. J. C. Reif, S. Hamrit, M. Heckenberger, et al., "Trends in genetic diversity among European maize cultivars and their parental components during the past 50 years," Theoretical and Applied Genetics, vol. 111, no. 5, pp. 838–845, 2005.
18. D. Hoisington, M. Khairallah, T. Reeves, et al., "Plant genetic resources: what can they contribute toward increased crop productivity?" Proceedings of the National Academy of Sciences of the United States of America, vol. 96, no. 11, pp. 5937–5943, 1999.
19. Y. Vigouroux, S. Mitchell, Y. Matsuoka, et al., "An analysis of genetic diversity across the maize genome using microsatellites," Genetics, vol. 169, no. 3, pp. 1617–1630, 2005.
20. S. I. Wright, I. V. Bi, S. C. Schroeder, et al., "Evolution: the effects of artificial selection on the maize genome," Science, vol. 308, no. 5726, pp. 1310–1314, 2005.

21. J. Dvořák, M.-C. Luo, and Z.-L. Yang, "Restriction fragment length polymorphism and divergence in the genomic regions of high and low recombination in self-fertilizing and cross- fertilizing Aegilops species,"Genetics, vol. 148, no. 1, pp. 423–434, 1998.
22. M. T. Hamblin and C. F. Aquadro, "DNA sequence variation and the recombinational landscape inDrosophila pseudoobscura: a study of the second chromosome," Genetics, vol. 153, no. 2, pp. 859–869, 1999.
23. B. S. Gaut, M. Le Thierry d›Ennequin, A. S. Peek, and M. C. Sawkins, "Maize as a model for the evolution of plant nuclear genomes," Proceedings of the National Academy of Sciences of the United States of America, vol. 97, no. 13, pp. 7008–7015, 2000.
24. R.-L. Wang, A. Stec, J. Hey, L. Lukens, and J. Doebley, "The limits of selection during maize domestication,"Nature, vol. 398, no. 6724, pp. 236–239, 1999.
25. J. F. Doebley, B. S. Gaut, and B. D. Smith, "The molecular genetics of crop domestication," Cell, vol. 127, no. 7, pp. 1309–1321, 2006.
26. A. L. Weber, W. H. Briggs, J. Rucker, et al., "The genetic architecture of complex traits in teosinte (Zea mays ssp. parviglumis): new evidence from association mapping," Genetics, vol. 180, no. 2, pp. 1221–1232, 2008.
27. J. Franco, J. Crossa, S. Taba, and H. Shands, "A sampling strategy for conserving genetic diversity when forming core subsets," Crop Science, vol. 45, no. 3, pp. 1035–1044, 2005.
28. J. Franco, J. Crossa, M. L. Warburton, and S. Taba, "Sampling strategies for conserving maize diversity when forming core subsets using genetic markers," Crop Science, vol. 46, no. 2, pp. 854–864, 2006.
29. C. E. Harjes, T. R. Rocheford, L. Bai, et al., "Natural genetic variation in lycopene epsilon cyclase tapped for maize biofortification," Science, vol. 319, no. 5861, pp. 330–333, 2008.
30. S. J. Emrich, L. Li, T.-J. Wen, et al., "Nearly identical paralogs: implications for maize (Zea mays L.) genome evolution," Genetics, vol. 175, no. 1, pp. 429–439, 2007.

31. L. Comai, K. Young, B. J. Till, et al., "Efficient discovery of DNA polymorphisms in natural populations by Ecotilling," Plant Journal, vol. 37, no. 5, pp. 778–786, 2004.
32. M. Gore, P. Bradbury, R. Hogers, et al., "Evaluation of target preparation methods for single-feature polymorphism detection in large complex plant genomes," Crop Science, vol. 47, pp. S135–S148, 2007.
33. W. B. Barbazuk, S. J. Emrich, H. D. Chen, L. Li, and P. S. Schnable, "SNP discovery via 454 transcriptome sequencing," Plant Journal, vol. 51, no. 5, pp. 910–918, 2007.
34. J. Yu and E. S. Buckler, "Genetic association mapping and genome organization of maize," Current Opinion in Biotechnology, vol. 17, no. 2, pp. 155–160, 2006.
35. E. S. Jones, H. Sullivan, D. Bhattramakki, and J. S. C. Smith, "A comparison of simple sequence repeat and single nucleotide polymorphism marker technologies for the genotypic analysis of maize (Zea mays L.),"Theoretical and Applied Genetics, vol. 115, no. 3, pp. 361–371, 2007.
36. R. A. Emerson, G. W. Beadle, and A. C. Fraser, "A summary of linkage studies in maize," Cornell University Agricultural Experiment Station Memoirs, vol. 180, pp. 1–83, 1935.
37. E. H. Coe, D. A. Hoisington, and M. G. Neuffer, "Linkage map of corn (maize) (Zea mays L.)," Maize Genetics Cooperative Newsletter, vol. 61, pp. 116–147, 1987.
38. B. Burr, F. A. Burr, K. H. Thompson, M. C. Albertson, and C. W. Stuber, "Gene mapping with recombinant inbreds in maize," Genetics, vol. 118, no. 3, pp. 519–526, 1988.
39. M. Lee, N. Sharopova, W. D. Beavis, et al., "Expanding the genetic map of maize with the intermatedB73×Mo17 (IBM) population," Plant Molecular Biology, vol. 48, no. 5-6, pp. 453–461, 2002.
40. W. D. Beavis and D. Grant, "A linkage map based on information from 4 F2 populations of maize (Zea mays L.)," Theoretical and Applied Genetics, vol. 82, pp. 636–644, 1991.
41. J. M. Gardiner, E. H. Coe, S. Melia-Hancock, D. A. Hoisington, and S. Chao, "Development of a core RFLP map in maize using

an immortalized F2 population," Genetics, vol. 134, no. 3, pp. 917–930, 1993.

42. G. L. Davis, M. D. McMullen, C. Baysdorfer, et al., "A maize map standard with sequenced core markers, grass genome reference points and 932 expressed sequence tagged sites (ESTs) in a 1736-locus map,"Genetics, vol. 152, no. 3, pp. 1137–1172, 1999.
43. M. Causse, S. Santoni, C. Damerval, et al., "A composite map of expressed sequences in maize," Genome, vol. 39, no. 2, pp. 418–432, 1996.
44. G. Taramino and S. Tingey, "Simple sequence repeats for germplasm analysis and mapping in maize,"Genome, vol. 39, no. 2, pp. 277–287, 1996.
45. G. Davis, E. Coe Jr., O. Heredia-Diaz, et al., "Toward a consensus maize map," in Maize Genetics Conference Abstract, vol. 36, p. 27, 1994.
46. M. Falque, L. Décousset, D. Dervins, et al., "Linkage mapping of 1454 new maize candidate gene loci,"Genetics, vol. 170, no. 4, pp. 1957–1966, 2005.
47. J. Gardiner, S. Schroeder, M. L. Polacco, et al., "Anchoring 93 971 maize expressed sequence tagged unigenes to the bacterial artificial chromosome contig map by two-dimensional overgo hybridization,"Plant Physiology, vol. 134, pp. 1317–1326, 2004.
48. F. Wei, E. Coe, W. Nelson, et al., "Physical and genetic structure of the maize genome reflects its complex evolutionary history," PLoS Genetics, vol. 3, no. 7, pp. 1254–1263, 2007.
49. K. C. Cone and E. H. Coe, "Genetic mapping and maps," in Handbook of Maize: Genetics and Genomics, J. L. Bennetzen and S. C. Hake, Eds., pp. 507–522, Springer, Science+Business Media LLC, New York, NY, USA, 2009.
50. C. W. Stuber, M. D. Edwards, and J. F. Wendel, "Molecular marker-facilitated investigations of quantitative trait loci in maize. II. Factors influencing yield and its component traits," Crop Science, vol. 27, pp. 639–648, 1987.
51. M. D. Edwards, T. Helentjaris, S. Wright, and C. W. Stuber, "Molecular-marker-facilitated investigations of quantitative trait loci in maize. 4. Analysis based on genome saturation

with isozyme and restriction fragment length polymorphism markers," Theoretical and Applied Genetics, vol. 83, no. 6-7, pp. 765–774, 1992.

52. C. W. Stuber, S. E. Lincoln, D. W. Wolff, T. Helentjaris, and E. S. Lander, "Identification of genetic factors contributing to heterosis in a hybrid from two elite maize inbred lines using molecular markers," Genetics, vol. 132, no. 3, pp. 823–839, 1992.
53. C. C. Schon, H. F. Utz, S. Groh, B. Truberg, S. Openshaw, and A. E. Melchinger, "Quantitative trait locus mapping based on resampling in a vast maize testcross experiment and its relevance to quantitative genetics for complex traits," Genetics, vol. 167, no. 1, pp. 485–498, 2004
54. C. C. Laurie, S. D. Chasalow, J. R. LeDeaux, et al., "The genetic architecture of response to long-term artificial selection for oil concentration in the maize kernel," Genetics, vol. 168, no. 4, pp. 2141–2155, 2004.
55. J. M. Ribaut, M. Bänziger, T. Setter, G. Edmeades, and D. Hoisington, "Genetic dissection of drought tolerance in maize: a case study," in Physiology and Biotechnology Integration for Plant Breeding, H. Nguyen and A. Blum, Eds., pp. 571–611, Marcel Dekker, New York, NY, USA, 2004.
56. M. C. Sawkins, A. D. Farmer, D. Hoisington, et al., "Comparative Map and Trait Viewer (CMTV): an integrated bioinformatic tool to construct consensus maps and compare QTL and functional genomics data across genomes and experiments," Plant Molecular Biology, vol. 56, no. 3, pp. 465–480, 2004.
57. M. Schaeffer, P. Byrne, and E. H. Coe Jr., "Consensus quantitative trait maps in maize: a database strategy,"Maydica, vol. 51, no. 2, pp. 357–367, 2006.
58. R. J. Wisser, P. J. Balint-Kurti, and R. J. Nelson, "The genetic architecture of disease resistance in maize: a synthesis of published studies," Phytopathology, vol. 96, no. 2, pp. 120–129, 2006.
59. R. Tuberosa and S. Salvi, "QTL for agronomic traits in maize production," in Handbook of Maize: Its Biology, J. L. Bennetzen and S. C. Hake, Eds., pp. 501–541, Springer, Science+Business Media LLC, London, UK, 2009.

60. B. C. Y. Collard, M. Z. Z. Jahufer, J. B. Brouwer, and E. C. K. Pang, "An introduction to markers, quantitative trait loci (QTL) mapping and marker-assisted selection for crop improvement: the basic concepts," Euphytica, vol. 142, no. 1-2, pp. 169–196, 2005.
61. Y. Xu and J. H. Crouch, "Marker-assisted selection in plant breeding: from publications to practice," Crop Science, vol. 48, no. 2, pp. 391–407, 2008.
62. Y. Xu, J. Wang, and J. H. Crouch, "Selective genotyping and pooled DNA analysis: An innovative use of an old concept," in Recognizing Past Achievement, Meeting Future Needs, Proceedings of the 5th International Crop Science Congress, Jeju, Korea, April 2008.
63. P. Landi, M. C. Sanguineti, S. Salvi, et al., "Validation and characterization of a major QTL affecting leaf ABA concentration in maize," Molecular Breeding, vol. 15, no. 3, pp. 291–303, 2005.
64. J. Doebley and A. Stec, "Inheritance of the morphological differences between maize and teosinte: comparison of results for two F2 populations," Genetics, vol. 134, no. 2, pp. 559–570, 1993.
65. J. Doebley, A. Stec, and C. Gustus, "Teosinte branched 1 and the origin of maize: evidence for epistasis and the evolution of dominance," Genetics, vol. 141, pp. 333–346, 1995.
66. J. Doebley, A. Stec, and L. Hubbard, "The evolution of apical dominance in maize," Nature, vol. 386, no. 6624, pp. 485–488, 1997.
67. J. Dorweiler, A. Stec, J. Kermicle, and J. Doebley, "Teosinte glume architecture 1: a genetic locus controlling a key step in maize evolution," Science, vol. 262, no. 5131, pp. 233–235, 1993.
68. H. Wang, T. Nussbaum-Wagler, B. Li, et al., "The origin of the naked grains of maize," Nature, vol. 436, no. 7051, pp. 714–719, 2005.
69. W. H. Briggs, M. D. McMullen, B. S. Gaut, and J. Doebley, "Linkage mapping of domestication loci in a large maize-teosinte backcross resource," Genetics, vol. 177, no. 3, pp. 1915–1928, 2007.

70. T. Dhliwayo, M. Lee, A. Ross, W. Woodman-Clikeman, and A. Hallauer, "Genetic mapping and analysis of panicle morphology in maize," in Proceedings of the Plant & Animal Genomes XIV Conference, p. 369, 2006.

71. J. M. Thornsberry, M. M. Goodman, J. Doebley, S. Kresovich, D. Nielsen, and E. S. Buckler, "Dwarf8polymorphisms associate with variation in flowering time," Nature Genetics, vol. 28, no. 3, pp. 286–289, 2001.

72. K. A. Palaisa, M. Morgante, M. Williams, and A. Rafalski, "Contrasting effects of selection on sequence diversity and linkage disequilibrium at two phytoene synthase loci," The Plant Cell, vol. 15, no. 8, pp. 1795–1806, 2003.

73. L. M. Wilson, S. R. Whitt, A. M. Ibáñezc, T. R. Rocheford, M. M. Goodman, and E. S. Buckler IV, "Dissection of maize kernel composition and starch production by candidate gene association," The Plant Cell, vol. 16, no. 10, pp. 2719–2733, 2004.

74. J. R. Andersen, T. Schrag, A. E. Melchinger, I. Zein, and T. Lübberstedt, "Validation of Dwarf8polymorphisms associated with flowering time in elite European inbred lines of maize (Zea mays L.),"Theoretical and Applied Genetics, vol. 111, no. 2, pp. 206–217, 2005.

75. S. J. Szalma, E. S. Buckler IV, M. E. Snook, and M. D. McMullen, "Association analysis of candidate genes for maysin and chlorogenic acid accumulation in maize silks," Theoretical and Applied Genetics, vol. 110, no. 7, pp. 1324–1333, 2005.

76. L. Camus-Kulandaivelu, J.-B. Veyrieras, D. Madur, et al., "Maize adaptation to temperate climate: relationship between population structure and polymorphism in the Dwarf8 gene," Genetics, vol. 172, no. 4, pp. 2449–2463, 2006.

77. S. Salvi, G. Sponza, M. Morgante, et al., "Conserved noncoding genomic sequences associated with a flowering-time quantitative trait locus in maize," Proceedings of the National Academy of Sciences of the United States of America, vol. 104, no. 27, pp. 11376–11381, 2007.

78. S. Ducrocq, D. Madur, J.-B. Veyrieras, et al., "Key impact ofVgt1 on flowering time adaptation in maize: evidence from association

mapping and ecogeographical information," Genetics, vol. 178, no. 4, pp. 2433–2437, 2008.

79. C. J. Hale, J. B. Hollick, E. S. Ersoz, J. Yu, and E. S. Buckler, "Applications of linkage disequilibrium and association mapping in maize," in Molecular Genetic Approaches to Maize Improvement, A. Kriz and B. Larkins, Eds., pp. 173–195, Springer, Berlin, Germany, 2008.

80. C. Zhu, M. Gore, E. S. Buckler, and J. Yu, "Status and prospects of association mapping in plants," Plant Genome, vol. 1, pp. 5–20, 2008.

81. M. I. Tenaillon, M. C. Sawkins, A. D. Long, R. L. Gaut, J. F. Doebley, and B. S. Gaut, "Patterns of DNA sequence polymorphism along chromosome 1 of maize (Zea mays ssp. mays L.)," Proceedings of the National Academy of Sciences of the United States of America, vol. 98, no. 16, pp. 9161–9166, 2001.

82. D. L. Remington, J. M. Thornsberry, Y. Matsuoka, et al., "Structure of linkage disequilibrium and phenotypic associations in the maize genome," Proceedings of the National Academy of Sciences of the United States of America, vol. 98, no. 20, pp. 11479–11484, 2001.

83. M. Jung, A. Ching, D. Bhattramakki, et al., "Linkage disequilibrium and sequence diversity in a 500-kbp region around the adh1 locus in elite maize germplasm," Theoretical and Applied Genetics, vol. 109, no. 4, pp. 681–689, 2004.

84. J. Yu, J. B. Holland, M. D. McMullen, and E. S. Buckler, "Genetic design and statistical power of nested association mapping in maize," Genetics, vol. 178, no. 1, pp. 539–551, 2008.

85. B. Stich, J. Yu, A. E. Melchinger, et al., "Power to detect higher-order epistatic interactions in a metabolic pathway using a new mapping strategy," Genetics, vol. 176, no. 1, pp. 563–570, 2007.

86. Y. Xu, Y. Lu, L. Yan, et al., "SNP chip-based genomewide scans for germplasm evaluation, marker-trait association analysis and development of a molecular breeding platform in maize," in Proceedings of 14th Australasian Plant Breeding Conference & 11th Congress of the Society for the Advancement of Breeding Research in Asia and Oceania, Cairns Convention Centre, Cairns, Australia, August 2009.

87. A. Rafalski and E. Ananiev, "Genetic diversity, linkage disequilibrium and association mapping," inHandbook of Maize: Genetics and Genomics, J. L. Bennetzen and S. C. Hake, Eds., pp. 201–219, Springer, Science+Business Media LLC, 2009.
88. A. H. Price, "Believe it or not, QTLs are accurate!," Trends in Plant Science, vol. 11, no. 5, pp. 213–216, 2006.
89. D. W. Podlich, C. R. Winkler, and M. Cooper, "Mapping as you go: an effective approach for marker-assisted selection of complex traits," Crop Science, vol. 44, no. 5, pp. 1560–1571, 2004.
90. S. M. Kaeppler, "Quantitative trait locus mapping using sets of near-isogenic lines: relative power comparisons and technical considerations," Theoretical and Applied Genetics, vol. 95, no. 3, pp. 384–392, 1997.
91. S. J. Szalma, B. M. Hostert, J. R. LeDeaux, C. W. Stuber, and J. B. Holland, "QTL mapping with near-isogenic lines in maize," Theoretical and Applied Genetics, vol. 114, no. 7, pp. 1211–1228, 2007.
92. G. I. Graham, D. W. Wolff, and C. W. Stuber, "Characterization of a yield quantitative trait locus on chromosome five of maize by fine mapping," Crop Science, vol. 37, no. 5, pp. 1601–1610, 1997.
93. L. E. Palmer, P. D. Rabinowicz, A. L. O›Shaughnessy, et al., "Maize genome sequencing by methylation filtration," Science, vol. 302, no. 5653, pp. 2115–2117, 2003.
94. C. A. Whitelaw, W. B. Barbazuk, G. Pertea, et al., "Enrichment of gene-coding sequences in maize by genome filtration," Science, vol. 302, no. 5653, pp. 2118–2120, 2003.
95. N. M. Springer, X. Xu, and W. B. Barbazuk, "Utility of different gene enrichment approaches toward identifying and sequencing the maize gene space," Plant Physiology, vol. 136, no. 2, pp. 3023–3033, 2004.
96. P. D. Rabinowicz and J. L. Bennetzen, "The maize genome as a model for efficient sequence analysis of large plant genomes," Current Opinion in Plant Biology, vol. 9, no. 2, pp. 149–156, 2006.
97. Y. Fu, S. J. Emrich, L. Guo, et al., "Quality assessment of maize assembled genomic islands (MAGIs) and large-sclae experimental

verification of predicted genes," Proceedings of the National Academy of Sciences of the United States of America, vol. 102, no. 34, pp. 12282–12287, 2005.

98. J. Emberton, J. Ma, Y. Yuan, P. SanMiguel, and J. L. Bennetzen, "Gene enrichment in maize with hypomethylated partial restriction (HMPR) libraries," Genome Research, vol. 15, no. 10, pp. 1441–1446, 2005.
99. Y. Yuan, P. J. SanMiguel, and J. L. Bennetzen, "Methylation-spanning linker libraries link gene-rich regions and identify epigenetic boundaries in Zea mays," Genome Research, vol. 12, no. 9, pp. 1345–1349, 2002.
100. J. Shendure and H. Ji, "Next-generation DNA sequencing," Nature Biotechnology, vol. 26, no. 10, pp. 1135–1145, 2008.
101. P. K. Gupta, "Single-molecule DNA sequencing technologies for future genomics research," Trends in Biotechnology, vol. 26, no. 11, pp. 602–611, 2008.
102. A. Kahvejian, J. Quackenbush, and J. F. Thompson, "What would you do if you could sequence everything?" Nature Biotechnology, vol. 26, no. 10, pp. 1125–1133, 2008.
103. H. Candela and S. Hake, "The art and design of genetic screens: maize," Nature Reviews Genetics, vol. 9, no. 3, pp. 192–203, 2008.
104. E. Bortiri, D. Jackson, and S. Hake, "Advances in maize genomics: the emergence of positional cloning,"Current Opinion in Plant Biology, vol. 9, no. 2, pp. 164–171, 2006.
105. S. Brunner, K. Fengler, M. Morgante, S. Tingey, and A. Rafalski, "Evolution of DNA sequence nonhomologies among maize inbreds," The Plant Cell, vol. 17, no. 2, pp. 343–360, 2005.
106. R. Song and J. Messing, "Gene expression of a gene family in maize based on noncollinear haplotypes,"Proceedings of the National Academy of Sciences of the United States of America, vol. 100, no. 15, pp. 9055–9060, 2003.
107. R. Bruggmann, A. K. Bharti, H. Gundlach, et al., "Uneven chromosome contraction and expansion in the maize genome," Genome Research, vol. 16, no. 10, pp. 1241–1251, 2006.

108. Z.-Y. Wang, X.-F. Sun, F. Wang, K.-X. Tang, and J.-R. Zhang, "Enhanced resistance of snowdrop lectin (Galanthus nivalis L. agglutinin)-expressing maize to Asian corn borer (Ostrinia furnacalis Guenée),"Journal of Integrative Plant Biology, vol. 47, no. 7, pp. 873–880, 2005.

109. K. F. Erhard Jr., J. L. Stonaker, S. E. Parkinson, J. P. Lim, C. J. Hale, and J. B. Hollick, "RNA polymerase IV functions in paramutation in Zea mays," Science, vol. 323, no. 5918, pp. 1201–1205, 2009.

110. E. Vollbrecht, P. S. Springer, L. Goh, E. S. Buckler IV, and R. Martienssen, "Architecture of floral branch systems in maize and related grasses," Nature, vol. 436, no. 7054, pp. 1119–1126, 2005.

111. K. Komatsu, M. Maekawa, S. Ujiie, et al., "LAX and SPA: major regulators of shoot branching in rice,"Proceedings of the National Academy of Sciences of the United States of America, vol. 100, no. 20, pp. 11765–11770, 2003.

112. P. Bommert, C. Lunde, J. Nardmann, et al., "thick tassel dwarf1 encodes a putative maize ortholog of theArabidopsis CLAVATA1 leucine-rich repeat receptor-like kinase," Development, vol. 132, no. 6, pp. 1235–1245, 2005.

113. E. Bortiri, G. Chuck, E. Vollbrecht, T. Rocheford, R. Martienssen, and S. Hake, "ramosa2 encodes a LATERAL ORGAN BOUNDARY domain protein that determines the fate of stem cells in branch meristems of maize," The Plant Cell, vol. 18, no. 3, pp. 574–585, 2006.

114. N. Satoh-Nagasawa, N. Nagasawa, S. Malcomber, H. Sakai, and D. Jackson, "A trehalose metabolic enzyme controls inflorescence architecture in maize," Nature, vol. 441, no. 7090, pp. 227–230, 2006.

115. A. Gallavotti, S. Barazesh, S. Malcomber, et al., "sparse inflorescence1 encodes a monocot-specific YUCCA-like gene required for vegetative and reproductive development in maize," Proceedings of the National Academy of Sciences of the United States of America, vol. 105, no. 39, pp. 15196–15201, 2008.

116. M. M. S. Evans, "The indeterminate gametophyte1 gene of maize encodes a LOB domain protein required for embryo sac and leaf development," The Plant Cell, vol. 19, no. 1, pp. 46–62, 2007.

117. Y. Barrière and O. Argillier, "Brown-midrib genes of maize: a review," Agronomie, vol. 13, no. 10, pp. 865–876, 1993.
118. Y. Barrière, C. Guillet, D. Goffner, and M. Pichon, "Genetic variation and breeding strategies for improved cell wall digestibility in annual forage crops: a review," Animal Research, vol. 52, no. 3, pp. 193–228, 2003.
119. T. Lübberstedt, I. Zein, J. R. Andersen, et al., "Development and application of functional markers in maize," Euphytica, vol. 146, no. 1-2, pp. 101–108, 2005.
120. C. Shi, A. Uzarowska, M. Ouzunova, M. Landbeck, G. Wenzel, and T. Lübberstedt, "Identification of candidate genes associated with cell wall digestibility and eQTL (expression quantitative trait loci) analysis in a Flint × Flint maize recombinant inbred line population," BMC Genomics, vol. 8, article 22, 2007.
121. W. Xiao, M. Xu, J. Zhao, F. Wang, J. Li, and J. Dai, "Genome-wide isolation and mapping of resistance gene analogs," Theoretical and Applied Genetics, vol. 113, pp. 63–72, 2006.
122. A. Sindhu, S. Chintamanani, A. S. Brandt, M. Zanis, S. R. Scofield, and G. S. Johal, "A guardian of grasses: specific origin and conservation of a unique disease-resistance gene in the grass lineage," Proceedings of the National Academy of Sciences of the United States of America, vol. 105, no. 5, pp. 1762–1767, 2008.
123. C. Lin, B. Shen, Z. Xu, T. G. Köllner, J. Degenhardt, and H. K. Dooner, "Characterization of the monoterpene synthase gene tps26, the ortholog of a gene induced by insect herbivory in maize," Plant Physiology, vol. 146, no. 3, pp. 940–951, 2008.
124. M. M. Herrmann, S. Pinto, J. Kluth, U. Wienand, and R. Lorbiecke, "The PTI1-like kinase ZmPti1a from maize (Zea mays L.) co-localizes with callose at the plasma membrane of pollen and facilitates a competitive advantage to the male gametophyte," BMC Plant Biology, vol. 6, article 22, 2006.
125. F. Hochholdinger, T.-J. Wen, R. Zimmermann, et al., "The maize (Zea mays L.) roothairless3 gene encodes a putative GPI-anchored, monocot-specific, COBRA-like protein that significantly affects grain yield," The Plant Journal, vol. 54, no. 5, pp. 888–898, 2008.

126. B. P. May, H. Liu, E. Vollbrecht, et al., "Maize-targeted mutagenesis: a knockout resource for maize,"Proceedings of the National Academy of Sciences of the United States of America, vol. 100, no. 20, pp. 11541–11546, 2003.
127. D. R. McCarty, A. M. Settles, M. Suzuki, et al., "Steady-state transposon mutagenesis in inbred maize," The Plant Journal, vol. 44, no. 1, pp. 52–61, 2005.
128. C. M. McCallum, L. Comai, E. A. Greene, and S. Henikoff, "Targeting induced local lesions IN genomes (TILLING) for plant functional genomics," Plant Physiology, vol. 123, no. 2, pp. 439–442, 2000.
129. B. J. Till, L. Comai, and S. abd Henikoff, "TILLING and EcoTILLING for crop improvement," in Genomics-Assisted Crop Improvement. Vol.1: Genomic Approaches and Platforms, R. K. Varshney and R. Tuberosa, Eds., pp. 333–350, Springer, Dordrecht, The Netherlands, 2007.
130. C. Weil and R. Monde, "TILLING and point mutation detection," in Handbook of Maize: Genetics and Genomics, J. L. Bennetzen and S. C. Hake, Eds., pp. 585–596, Springer, Science+Business Media LLC, London, UK, 2009.
131. D. R. McCarty and R. B. Meeley, "Transposon resources for forward and reverse genetics in maize," inHandbook of Maize: Genetics and Genomics, J. L. Bennetzen and S. C. Hake, Eds., pp. 561–584, Springer, Science+Business Media LLC, 2009.
132. A. L. Iniguez, J. Gardiner, M. Hogan, et al., "Antisense expression analysis in the maize transcriptome and microarray cross platform comparisons," in Maize Genetics Conference, 2006, Abstract 48, P2.
133. V. G. Cheung and R. S. Spielman, "The genetics of variation in gene expression," Nature Genetics, vol. 32, supplement, pp. 522–525, 2002.
134. R. C. Jansen and J.-P. Nap, "Genetical genomics: the added value from segregation," Trends in Genetics, vol. 17, no. 7, pp. 388–391, 2001.
135. R. A. Swanson-Wagner, Y. Jia, R. DeCook, L. A. Borsuk, D. Nettleton, and P. S. Schnable, "All possible modes of gene action are observed in a global comparison of gene expression

in a maize F1 hybrid and its inbred parents," Proceedings of the National Academy of Sciences of the United States of America, vol. 103, no. 18, pp. 6805–6810, 2006.

136. R. M. Stupar and N. M. Springer, "Cis-transcriptional variation in maize inbred lines B73 and Mo17 leads to additive expression patterns in the F1 hybrid," Genetics, vol. 173, no. 4, pp. 2199–2210, 2006.

137. N. M. Springer and R. M. Stupar, "Allele-specific expression patterns reveal biases and embryo-specific parent-of-origin effects in hybrid maize," The Plant Cell, vol. 19, no. 8, pp. 2391–2402, 2007.

138. R. M. Stupar, P. J. Hermanson, and N. M. Springer, "Nonadditive expression and parent-of-origin effects identified by microarray and allele-specific expression profiling of maize endosperm," Plant Physiology, vol. 145, no. 2, pp. 411–425, 2007.

139. R. M. Stupar, J. M. Gardiner, A. G. Oldre, W. J. Haun, V. L. Chandler, and N. M. Springer, "Gene expression analyses in maize inbreds and hybrids with varying levels of heterosis," BMC Plant Biology, vol. 8, article 33, 2008.

140. G. Pea, S. Ferron, L. Gianfranceschi, P. Krajewski, and M. Enrico Pè, "Gene expression non-additivity in immature ears of a heterotic F1 maize hybrid," Plant Science, vol. 174, no. 1, pp. 17–24, 2008.

141. N. Hoecker, B. Keller, N. Muthreich, et al., "Comparison of maize (Zea mays L.) F1-hybrid and parental inbred line primary root transcriptomes suggests organ-specific patterns of nonadditive gene expression and conserved expression trends," Genetics, vol. 179, no. 3, pp. 1275–1283, 2008.

142. S. Brenner, M. Johnson, J. Bridgham, et al., "Gene expression analysis by massively parallel signature sequencing (MPSS) on microbead arrays," Nature Biotechnology, vol. 18, no. 6, pp. 630–634, 2000.

143. M. Guo, S. Yang, M. Rupe, et al., "Genome-wide allele-specific expression analysis using Massively Parallel Signature Sequencing (MPSS) reveals cis- and trans-effects on gene expression in maize hybrid meristem tissue," Plant Molecular Biology, vol. 66, no. 5, pp. 551–563, 2008.

144. K. P. Kollipara, I. N. Saab, R. D. Wych, M. J. Lauer, and G. W. Singletary, "Expression profiling of reciprocal maize hybrids divergent for cold germination and desiccation tolerance," Plant Physiology, vol. 129, no. 3, pp. 974–992, 2002.

145. T. Umezawa, M. Fujita, Y. Fujita, K. Yamaguchi-Shinozaki, and K. Shinozaki, "Engineering drought tolerance in plants: discovering and tailoring genes to unlock the future," Current Opinion in Biotechnology, vol. 17, no. 2, pp. 113–122, 2006.

146. B. Valliyodan and H. T. Nguyen, "Understanding regulatory networks and engineering for enhanced drought tolerance in plants," Current Opinion in Plant Biology, vol. 9, no. 2, pp. 189–195, 2006.

147. N. Sreenivasulu, S. K. Sopory, and P. B. Kavi Kishor, "Deciphering the regulatory mechanisms of abiotic stress tolerance in plants by genomic approaches," Gene, vol. 388, no. 1-2, pp. 1–13, 2007.

148. J. Ma, D. J. Morrow, J. Fernandes, and V. Walbot, "Comparative profiling of the sense and antisense transcriptome of maize lines," Genome Biology, vol. 7, pp. R22.1–R22.18, 2006.

149. C. Calderon-Vazquez, E. Ibarra-Laclette, J. Caballero-Perez, and L. Herrera-Estrella, "Transcript profiling of Zea mays roots reveals gene responses to phosphate deficiency at the plant- and species-specific levels,"Journal of Experimental Botany, vol. 59, no. 9, pp. 2479–2497, 2008.

150. J. Fernandes, D. J. Morrow, P. Casati, and V. Walbot, "Distinctive transcriptome responses to adverse environmental conditions in Zea mays L.," Plant Biotechnology Journal, vol. 6, no. 8, pp. 782–798, 2008.

151. A. K. Shrawat and L. Horst, "Agrobacterium-mediated transformation of cereals: a promising approach crossing barriers," Plant Biotechnology Journal, vol. 4, no. 6, pp. 575–603, 2006.

152. P. Meyer and H. Saedler, "Homology-dependent gene silencing in plants," Annual Review of Plant Physiology and Plant Molecular Biology, vol. 47, no. 1, pp. 23–48, 1996.

153. H. Shou, B. R. Frame, S. A. Whitham, and K. Wang, "Assessment of transgenic maize events produced by particle bombardment

or Agrobacterium-mediated transformation," Molecular Breeding, vol. 13, no. 2, pp. 201–208, 2004.

154. Y. Hiei, S. Ohta, T. Komari, and T. Kumashiro, "Efficient transformation of rice (Oryza sativa L.) mediated by Agrobacterium and sequence analysis of the boundaries of the T-DNA," The Plant Journal, vol. 6, no. 2, pp. 271–282, 1994.
155. Y. Ishida, H. Saito, S. Ohta, Y. Hiei, T. Komari, and T. Kumashiro, "High efficiency transformation of maize (Zea mays L.) mediated by Agrobacterium tumefaciens," Nature Biotechnology, vol. 14, no. 6, pp. 745–750, 1996.
156. D. Negrotto, M. Jolley, S. Beer, A. R. Wenck, and G. Hansen, "The use of phosphomannose-isomerase as a selectable marker to recover transgenic maize plants (Zea mays L.) via Agrobacterium transformation,"Plant Cell Reports, vol. 19, no. 8, pp. 798–803, 2000.
157. Z.-Y. Zhao, W. Gu, T. Cai, et al., "High throughput genetic transformation mediated by Agrobacterium tumefaciens in maize," Molecular Breeding, vol. 8, no. 4, pp. 323–333, 2001.
158. B. R. Frame, H. Shou, R. K. Chikwamba, et al., "Agrobacterium tumefaciens-mediated transformation of maize embryos using a standard binary vector system," Plant Physiology, vol. 129, no. 1, pp. 13–22, 2002.
159. J. M. Vega, W. Yu, A. R. Kennon, X. Chen, and Z. J. Zhang, "Improvement of Agrobacterium-mediated transformation in Hi-II maize (Zea mays) using standard binary vectors," Plant Cell Reports, vol. 27, no. 2, pp. 297–305, 2008.
160. C. L. Armstrong and J. R. Rout, "A novel Agrobacterium-mediated plant transformation method," Int. Patent no. W001/09302 A2, 2001.
161. W. Zhang, S. Subbarao, P. Addae, et al., "Cre/lox-mediated marker gene excision in transgenic maize (Zea mays L.) plants," Theoretical and Applied Genetics, vol. 107, no. 7, pp. 1157–1168, 2003.
162. Y. Ishida, Y. Hiei, and T. Komari, "Agrobacterium-mediated transformation of maize," Nature Protocols, vol. 2, no. 7, pp. 1614–1621, 2007.

163. J. M. Vega, W. Yu, A. R. Kennon, X. Chen, and Z. J. Zhang, "Improvement of Agrobacterium-mediated transformation in Hi-II maize (Zea mays) using standard binary vectors," Plant Cell Reports, vol. 27, no. 2, pp. 297–305, 2008.
164. A. Yang, C. He, K. Zhang, and J. Zhang, "Improvement of Agrobacterium-mediated transformation of embryogenic calluses from maize elite inbred lines," In Vitro Cellular and Developmental Biology—Plant, vol. 42, no. 3, pp. 215–219, 2006.
165. R. L. Roberts, M. Metz, D. E. Monks, M. L. Mullaney, T. Hall, and E. W. Nester, "Purine synthesis and increased Agrobacterium tumefaciens transformation of yeast and plants," Proceedings of the National Academy of Sciences of the United States of America, vol. 100, no. 11, pp. 6634–6639, 2003.
166. W. Gordon-Kamm, B. P. Dilkes, K. Lowe, et al., "Stimulation of the cell cycle and maize transformation by disruption of the plant retinoblastoma pathway," Proceedings of the National Academy of Sciences of the United States of America, vol. 99, no. 18, pp. 11975–11980, 2002.
167. L.-Y. Lee, M. E. Kononov, B. Bassuner, B. R. Frame, K. Wang, and S. B. Gelvin, "Novel plant transformation vectors containing the superpromoter," Plant Physiology, vol. 145, no. 4, pp. 1294–1300, 2007.
168. W. Yu, F. Han, Z. Gao, J. M. Vega, and J. A. Birchler, "Construction and behavior of engineered minichromosomes in maize," Proceedings of the National Academy of Sciences of the United States of America, vol. 104, no. 21, pp. 8924–8929, 2007.
169. Y. Ishida, H. Saito, Y. Hiei, and T. Komari, "Improved protocol for transformation of maize (Zea mays L.) mediated by Agrobacterium tumefaciens," Plant Biotechnology, vol. 14, pp. 745–750, 2003.
170. X. Huang and Z. Wei, "Successful Agrobacterium-mediated genetic transformation of maize elite inbred lines," Plant Cell, Tissue and Organ Culture, vol. 83, no. 2, pp. 187–200, 2005.
171. B. R. Frame, J. M. McMurray, T. M. Fonger, et al., "Improved Agrobacterium-mediated transformation of three maize inbred lines using MS salts," Plant Cell Reports, vol. 25, no. 10, pp. 1024–1034, 2006.

172. E. Lupotto, E. Conti, A. Reali, C. Lanzanova, E. Baldoni, and L. Allegri, "Improving in vitro culture and regeneration conditions for Agrobacterium-mediated maize transformation," Maydica, vol. 49, no. 1, pp. 21–29, 2004.

173. M. E. Horn, R. L. Harkey, A. K. Vinas, C. F. Drees, D. K. Barker, and J. R. Lane, "Use of Hi II-elite inbred hybrids in Agrobacterium-based transformation of maize," In Vitro Cellular and Developmental Biology—Plant, vol. 42, no. 4, pp. 359–366, 2006.

174. M. R. Willman, S. M. Schroll, and T. K. Hodges, "Inheritance of somatic embryogenesis and plantlet regeneration from primary (type 1) callus in maize," In Vitro Cellular & Developmental Biology—Plant, vol. 25, pp. 95–100, 1989.

175. A. Carlson and H. Kaeppler, "Genetic analysis of somatic embryogenesis in maize inbred A188," in Maize Genetics Conference Abstract, vol. 48, p. 97, 2006.

176. B. A. Lowe, M. M. Way, J. M. Kumpf, et al., "Marker assisted breeding for transformability in maize,"Molecular Breeding, vol. 18, no. 3, pp. 229–239, 2006.

177. S. M. Cocciolone, D. Nettleton, M. E. Snook, and T. Peterson, "Transformation of maize with the p1 transcription factor directs production of silk maysin, a corn earworm resistance factor, in concordance with a hierarchy of floral organ pigmentation," Plant Biotechnology Journal, vol. 3, no. 2, pp. 225–235, 2005.

178. S. Naqvi, C. Zhu, G. Farre, et al., "Transgenic multivitamin corn through biofortification of endosperm with three vitamins representing three distinct metabolic pathways," Proceedings of the National Academy of Sciences of the United States of America, vol. 106, no. 19, pp. 7762–7767, 2009.

179. S. Ohta, Y. Ishida, and S. Usami, "Expression of cold-tolerant pyruvate, orthophosphate dikinase cDNA, and heterotetramer formation in transgenic maize plants," Transgenic Research, vol. 13, no. 5, pp. 475–485, 2004.

180. H. Shou, P. Bordallo, J.-B. Fan, et al., "Expression of an active tabacco mitogen-activated protein kinase kinase kinase enhances freezing tolerance in transgenic maize," Proceedings of the National Academy of Sciences of the United States of America, vol. 101, no. 9, pp. 3298–3303, 2004.

181. J. Sheen, "Protein phosphatase activity is required for light-inducible gene expression in maize," The EMBO Journal, vol. 12, no. 9, pp. 3497–3505, 1993.
182. H. Shou, P. Bordallo, and K. Wang, "Expression of the Nicotiana protein kinase (NPK1) enhanced drought tolerance in transgenic maize," Journal of Experimental Botany, vol. 55, no. 399, pp. 1013–1019, 2004.
183. R. Quan, M. Shang, H. Zhang, Y. Zhao, and J. Zhang, "Improved chilling tolerance by transformation with betA gene for the enhancement of glycinebetaine synthesis in maize," Plant Science, vol. 166, no. 1, pp. 141–149, 2004.
184. X.-Y. Yin, A.-F. Yang, K.-W. Zhang, and J.-R. Zhang, "Production and analysis of transgenic maize with improved salt tolerance by the introduction of AtNHX1 gene," Acta Botanica Sinica, vol. 46, no. 7, pp. 854–861, 2004.
185. P. Castiglioni, D. Warner, R. J. Bensen, et al., "Bacterial RNA chaperones confer abiotic stress tolerance in plants and improved grain yield in maize under water-limited conditions," Plant Physiology, vol. 147, no. 2, pp. 446–455, 2008.
186. D. E. Nelson, P. P. Repetti, T. R. Adams, et al., "Plant nuclear factor Y (NF-Y) B subunits confer drought tolerance and lead to improved corn yields on water-limited acres," Proceedings of the National Academy of Sciences of the United States of America, vol. 104, no. 42, pp. 16450–16455, 2007.
187. S. Ortiz-García, E. Ezcurra, B. Schoel, F. Acevedo, J. Soberón, and A. A. Snow, "Absence of detectable transgenes in local landraces of maize in Oaxaca, Mexico (2003-2004)," Proceedings of the National Academy of Sciences of the United States of America, vol. 102, no. 35, pp. 12338–12343, 2005.
188. G. Brookes and P. Barfoot, "GM crops: the first ten years—global socio-economic and environmental impacts," ISAAA Briefs 36, 2006.
189. C. F. Chilcutt and B. E. Tabashnik, "Contamination of refuges by Bacillus thuringiensis toxin genes from transgenic maize," Proceedings of the National Academy of Sciences of the United States of America, vol. 101, no. 20, pp. 7526–7529, 2004.

190. S. L. Bates, J.-Z. Zhao, R. T. Roush, and A. M. Shelton, "Insect resistance management in GM crops: past, present and future," Nature Biotechnology, vol. 23, no. 1, pp. 57–62, 2005.

191. D. Quist and I. H. Chapela, "Transgenic DNA introgressed into traditional maize landraces in Oaxaca, Mexico," Nature, vol. 414, no. 6863, pp. 541–543, 2001.

192. D. Quist and I. H. Chapela, "Communications arising: response to criticism of transgenic DNA introgressed into traditional maize landraces in Oaxaca, Mexico," Nature, vol. 416, pp. 602–603, 2002.

193. D. A. Cleveland, D. Soleri, F. Aragón Cuevas, J. Crossa, and P. Gepts, "Detecting (trans)gene flow to landraces in centers of crop origin: lessons from the case of maize in Mexico," Environmental Biosafety Research, vol. 4, no. 4, pp. 197–208, 2005.

194. S. Ortiz-García, E. Ezcurra, B. Schoel, F. Acevedo, J. Soberón, and A. A. Snow, "Reply to Cleveland et al.›s "Detecting (trans) gene flow to landraces in centers of crop origin: lessons from the case of maize in Mexico"," Environmental Biosafety Research, vol. 4, no. 4, pp. 209–215, 2005.

195. R. M. Welch and R. D. Graham, "A new paradigm for world agriculture: meeting human needs. Productive, sustainable, nutritious," Field Crops Research, vol. 60, no. 1-2, pp. 1–10, 1999.

196. D. DellaPenna, "Nutritional genomics: manipulating plant micronutrients to improve human health,"Science, vol. 285, no. 5426, pp. 375–379, 1999.

197. H. Brinch-Pedersen, L. D. Sørensen, and P.B. Holm, "Engineering crop plants: getting a handle on phosphate," Trends in Plant Science, vol. 7, no. 3, pp. 118–125, 2002.

198. E. Graf, K. L. Empson, and J. W. Eaton, "Phytic acid: a natural antioxidant," Journal of Biological Chemistry, vol. 262, no. 24, pp. 11647–11650, 1987.

199. R. P. Glahn, G. M. Wortley, P. K. South, and D. D. Miller, "Inhibition of iron uptake by phytic acid, tannic acid, and ZnCl2: studies using an in vitro digestion/Caco-2 cell model," Journal of Agricultural and Food Chemistry, vol. 50, no. 2, pp. 390–395, 2002.

200. V. Raboy, K. A. Young, J. A. Dorsch, and A. Cook, "Genetics and breeding of seed phosphorus and phytic acid," Journal of Plant Physiology, vol. 158, no. 4, pp. 489–497, 2001.
201. R. Pilu, D. Panzeri, G. Gavazzi, S. K. Rasmussen, G. Consonni, and E. Nielsen, "Phenotypic, genetic and molecular characterization of a maize low phytic acid mutant (lpa241)," Theoretical and Applied Genetics, vol. 107, no. 6, pp. 980–987, 2003.
202. J. Shi, H. Wang, Y. Wu, J. Hazebroek, R. B. Meeley, and D. S. Ertl, "The maize low-phytic acid mutant lpa2 is caused by mutation in an inositol phosphate kinase gene," Plant Physiology, vol. 131, no. 2, pp. 507–515, 2003.
203. J. Shi, H. Wang, K. Schellin, et al., "Embryo-specific silencing of a transporter reduces phytic acid content of maize and soybean seeds," Nature Biotechnology, vol. 25, no. 8, pp. 930–937, 2007.
204. G. Drakakaki, S. Marcel, R. P. Glahn, et al., "Endosperm-specific co-expression of recombinant soybean ferritin and Aspergillus phytase in maize results in significant increases in the levels of bioavailable iron,"Plant Molecular Biology, vol. 59, no. 6, pp. 869–880, 2005.
205. R. S. Chauhan, "Bioinformatics approach toward identification of candidate genes for zinc and iron transporters in maize," Current Science, vol. 91, no. 4, pp. 510–515, 2006.
206. M. Banziger and J. Long, "The potential for increasing the iron and zinc density of maize through plant-breeding," Food and Nutrition Bulletin, vol. 21, no. 4, pp. 397–400, 2000.
207. J. K. Long, M. Bänziger, and M. E. Smith, "Diallel analysis of grain iron and zinc density in southern African-adapted maize inbreds," Crop Science, vol. 44, no. 6, pp. 2019–2026, 2004.
208. B. C. Gibbon and B. A. Larkins, "Molecular genetic approaches to developing quality protein maize,"Trends in Genetics, vol. 21, no. 4, pp. 227–233, 2005
209. National Research Council, Quality Protein Maize, National Academy Press, Washington, DC, USA, 1988.
210. B. G. Hunter, M. K. Beatty, G. W. Singletary, et al., "Maize opaque endosperm mutations create extensive changes in patterns of gene expression," The Plant Cell, vol. 14, no. 10, pp. 2591–2612, 2002.

211. M. Bjarnason and S. K. Vasal, "Breeding of quality protein maize (QPM)," Plant Breeding Reviews, vol. 9, pp. 181–216, 1992.
212. A. F. Krivanek, H. De Groote, N. S. Gunaratna, A. O. Diallo, and D. Friesen, "Breeding and disseminating quality protein maize (QPM) for Africa," African Journal of Biotechnology, vol. 6, no. 4, pp. 312–324, 2007.
213. S. Huang, A. Frizzi, C. A. Florida, D. E. Kruger, and M. H. Luethy, "High lysine and high tryptophan transgenic maize resulting from the reduction of both 19- and 22-kD α-zeins," Plant Molecular Biology, vol. 61, no. 3, pp. 525–535, 2006.
214. G. Segal, R. Song, and J. Messing, "A new opaque variant of maize by a single dominant RNA-interference-inducing transgene," Genetics, vol. 165, no. 1, pp. 387–397, 2003.
215. N. M. Houmard, J. L. Mainville, C. P. Bonin, S. Huang, M. H. Luethy, and T. M. Malvar, "High-lysine corn generated by endosperm-specific suppression of lysine catabolism using RNAi," Plant Biotechnology Journal, vol. 5, no. 5, pp. 605–614, 2007.
216. Q. Rascon-Cruz, S. Sinagawa-Garcia, J. A. Osuna-Castro, N. Bohorova, and O. Paredes-Lopez, "Accumulation, assembly, and digestibility of amarantin expressed in transgenic tropical maize,"Theoretical and Applied Genetics, vol. 108, no. 2, pp. 335–342, 2004.
217. O. V. Brenna and N. Berardo, "Application of near-infrared reflectance spectroscopy (NIRS) to the evaluation of carotenoids content in maize," Journal of Agricultural and Food Chemistry, vol. 52, no. 18, pp. 5577–5582, 2004.
218. J. A. Howe and S. A. Tanumihardjo, "Carotenoid-biofortified maize maintains adequate vitamin A status in Mongolian gerbils," Journal of Nutrition, vol. 136, no. 10, pp. 2562–2567, 2006.
219. R. M. Welch and R. D. Graham, "Breeding for micronutrients in staple food crops from a human nutrition perspective," Journal of Experimental Botany, vol. 55, no. 396, pp. 353–364, 2004.
220. C. O. Egesel, J. C. Wong, R. J. Lambert, and T. R. Rocheford, "Combining ability of maize inbreds for carotenoids and tocopherols," Crop Science, vol. 43, no. 3, pp. 818–823, 2003.

221. A. Menkir, W. Liu, W. S. White, B. Maziya-Dixon, and T. Rocheford, "Carotenoid diversity in tropical-adapted yellow maize inbred lines," Food Chemistry, vol. 109, no. 3, pp. 521–529, 2008.
222. B. Maziya-Dixon, J. G. Kling, A. Menkir, and A. Dixon, "Genetic variation in total carotene, iron, and zinc contents of maize and cassava genotypes," Food and Nutrition Bulletin, vol. 21, no. 4, pp. 419–422, 2000.
223. B. Buckner, P. S. Miguel, D. Janick-Buckner, and J. L. Bennetzen, "The y1 gene of maize codes for phytoene synthase," Genetics, vol. 143, no. 1, pp. 479–488, 1996.
224. M. Cervantes-Cervantes, V. Gallagher, C. Zhu, and E. Wurtzel, "Maize cDNAs expressed in endosperm encode functional farnesyl diphophate synthase with geranylgeranyl diphosphate synthase activity," Plant Physiology, pp. 220–231, 2006.
225. J. C. Wong, R. J. Lambert, E. T. Wurtzel, and T. R. Rocheford, "QTL and candidate genes phytoene synthase and ζ -carotene desaturase associated with the accumulation of carotenoids in maize," Theoretical and Applied Genetics, vol. 108, no. 2, pp. 349–359, 2004.
226. S. Chander, Y. Q. Guo, X. H. Yang, et al., "Using molecular markers to identify two major loci controlling carotenoid contents in maize grain," Theoretical and Applied Genetics, vol. 116, no. 2, pp. 223–233, 2008.
227. P. D. Fraser and P. M. Bramley, "The biosynthesis and nutritional uses of carotenoids," Progress in Lipid Research, vol. 43, no. 3, pp. 228–265, 2004.
228. H. Fu and H. K. Dooner, "Intraspecific violation of genetic colinearity and its implications in maize,"Proceedings of the National Academy of Sciences of the United States of America, vol. 99, no. 14, pp. 9573–9578, 2002.
229. Y. Xu, "Developing marker-assisted selection strategies for breeding hybrid rice," Plant Breeding Reviews, vol. 23, pp. 73–174, 2003.
230. W. S. Niebur, J. A. Rafalski, O. S. Smith, and M. Cooper, "Applications of genomics technologies to enhance rate of genetic progress for yield of maize within a commercial breeding program," in New Directions for a Diverse Planet: Proceedings

of the 4th International Crop Science Congress, T. Fisher, Ed., Brisbane, Australia, September-October 2004.

231. M. Cooper, O. S. Smith, G. Graham, L. Arthur, L. Feng, and D. W. Podlich, "Genomics, genetics, and plant breeding: a private sector perspective," Crop Science, vol. 44, no. 6, pp. 1907–1913, 2004.

232. Y. Xu, S. R. McCouch, and Q. Zhang, "How can we use genomics to improve cereals with rice as a reference genome?" Plant Molecular Biology, vol. 59, no. 1, pp. 7–26, 2005.

233. T. M. Crosbie, S. R. Eathington, G. R. Johnson, et al., "Plant breeding: past, present, and future," in Plant Breeding: The Arnel R. Hallauer International Symposium, K. R. Lamkey and M. Lee, Eds., pp. 3–50, Blackwell, Ames, Iowa, USA, 2006.

234. S. R. Eathington, T. M. Crosbie, M. D. Edwards, R. S. Reiter, and J. K. Bull, "Molecular markers in a commercial breeding program," Crop Science, vol. 47, supplement 3, pp. S154–S163, 2007.

235. R. Bernardo, "Molecular markers and selection for complex traits in plants: learning from the last 20 years," Crop Science, vol. 48, no. 5, pp. 1649–1664, 2008.

236. M. Ragot, G. Gay, J.-P. Muller, and J. Durovray, "Efficient selection for the adaptation to the environment through QTL mapping and manipulation in maize," in Molecular Approaches for the Genetic Improvement of Cereals for Stable Production in Water-Limited Environments, J. M. Ribaut and D. Poland, Eds., pp. 128–130, CIMMYT, Mexico City, Mexico, 2000.

237. S. R. Eathington, "Practical applications of molecular technology in the development of commercial maize hybrids," in Proceedings of the 60th Annual Corn and Sorghum Seed Research Conferences, American Seed Trade Association, 2005.

238. M. Ragot and M. Lee, "Marker-assisted selection in maize: current status, potential, limitations and perspertives from the private and public sectors," in Marker-Assisted Selection, Current Status and Future Perspectives in Crops, Livestock, Forestry and Fish, E. P. Guimarães, J. Ruane, B. D. Scherf, A. Sonnino, and J. D. Dargie, Eds., pp. 117–150, Food and Agriculture Organization of the Unites Nations, Rome, Italy, 2007.

239. C. W. Stuber, M. Polacco, and M. L. Senior, "Synergy of empirical breeding, marker-assisted selection, and genomics to increase crop yield potential," Crop Science, vol. 39, no. 6, pp. 1571–1583, 1999.

240. J. M. Ribaut, M. Bänziger, J. Betran, et al., "Use of molecular markers in plant breeding: drought tolerance improvement in tropical maize," in Quantitative Genetics, Genomics, and Plant Breeding, M. S. Kang, Ed., pp. 85–99, CABI Publishing, 2002.

241. J.-M. Ribaut and M. Ragot, "Marker-assisted selection to improve drought adaptation in maize: the backcross approach, perspectives, limitations, and alternatives," Journal of Experimental Botany, vol. 58, no. 2, pp. 351–360, 2007.

242. R. Tuberosa and S. Salvi, "Genomics-based approaches to improve drought tolerance of crops," Trends in Plant Science, vol. 11, no. 8, pp. 405–412, 2006.

243. K. K. Jena and D. J. Mackill, "Molecular markers and their use in marker-assisted selection in rice," Crop Science, vol. 48, no. 4, pp. 1266–1276, 2008.

244. K. Xu, X. Xu, T. Fukao, et al., "Sub1A is an ethylene-response-factor-like gene that confers submergence tolerance to rice," Nature, vol. 442, no. 7103, pp. 705–708, 2006.

245. R. Fraley, Presentation at Monsanto European Investor Day, November 2006, http://www.monsanto.com.

246. J. Xiao, J. Li, L. Yuan, and S. D. Tanksley, "Dominance is the major genetic basis of heterosis in rice as revealed by QTL analysis using molecular markers," Genetics, vol. 140, no. 2, pp. 745–754, 1995.

247. S. B. Yu, J. X. Li, C. G. Xu, et al., "Importance of epistasis as the genetic basis of heterosis in an elite rice hybrid," Proceedings of the National Academy of Sciences of the United States of America, vol. 94, no. 17, pp. 9226–9231, 1997.

248. Z.-K. Li, L. J. Luo, H. W. Mei, et al., "Overdominant epistatic loci are the primary genetic basis of inbreeding depression and heterosis in rice. I. Biomass and grain yield," Genetics, vol. 158, no. 4, pp. 1737–1753, 2001.

249. L. J. Luo, Z.-K. Li, H. W. Mei, et al., "Overdominant epistatic loci are the primary genetic basis of inbreeding depression and

heterosis in rice. II. Grain yield components," Genetics, vol. 158, no. 4, pp. 1755–1771, 2001.

250. W. F. Tracy and M. A. Chandler, "The historical and biological basis of the concept of heterotic patterns in Corn Belt Dent maize," in Plant Breeding: The Arnel R. Hallauer International Symposium, K. R. Lamkey and M. Lee, Eds., pp. 219–233, Blackwell, Ames, Iowa, USA, 2006.

251. W. B. Bruce, G. O. Edmeades, and T. C. Barker, "Molecular and physiological approaches to maize improvement for drought tolerance," Journal of Experimental Botany, vol. 53, no. 366, pp. 13–25, 2002.

252. D. M. O›Sullivan, D. Edwards, and K. J. Edwards, "Maize genomics," AgBiotechNet, vol. 2, pp. 1–6, 1999.

253. J. C. Cushman and H. J. Bohnert, "Genomic approaches to plant stress tolerance," Current Opinion in Plant Biology, vol. 3, no. 2, pp. 117–124, 2000.

254. C. W. Stuber, "Mapping and manipulating quantitative traits in maize," Trends in Genetics, vol. 11, no. 12, pp. 477–481, 1995.

255. S. R. Eathington, J. W. Dudley, and G. K. Rufener II, "Usefulness of marker-QTL associations in early generation selection," Crop Science, vol. 37, no. 6, pp. 1686–1693, 1997.

256. A. Bouchez, F. Hospital, M. Causse, A. Gallais, and A. Charcosset, "Marker-assisted introgression of favorable alleles at quantitative trait loci between maize elite lines," Genetics, vol. 162, no. 4, pp. 1945–1959, 2002.

257. J. C. Ho, S. R. McCouch, and M. E. Smith, "Improvement of hybrid yield by advanced backcross QTL analysis in elite maize," Theoretical and Applied Genetics, vol. 105, no. 2-3, pp. 440–448, 2002.

258. G. G. Yousef and J. A. Juvik, "Enhancement of seedling emergence in sweet corn by marker-assisted backcrossing of beneficial QTL," Crop Science, vol. 42, no. 1, pp. 96–104, 2002.

259. J. H. Xia and Y. L. Zheng, "Molecular marker-assisted backcross breeding of maize Rf3 NIL and its efficient analysis," Acta Agronomic Sinica, vol. 28, pp. 339–344, 2002.

260. M. C. Willcox, M. M. Khairallah, D. Bergvinson, et al., "Selection for resistance to southwestern corn borer using marker-assisted and conventional backcrossing," Crop Science, vol. 42, no. 5, pp. 1516–1528, 2002.

261. S. A. Flint-Garcia, L. L. Darrah, M. D. McMullen, and B. E. Hibbard, "Phenotypic versus marker-assisted selection for stalk strength and second-generation European corn borer resistance in maize," Theoretical and Applied Genetics, vol. 107, no. 7, pp. 1331–1336, 2003.

262. R. Babu, S. K. Nair, A. Kumar, et al., "Two-generation marker-aided backcrossing for rapid conversion of normal maize lines to quality protein maize (QPM)," Theoretical and Applied Genetics, vol. 111, no. 5, pp. 888–897, 2005.

263. J. W. Danson, M. Mbogori, M. Kimani, M. Lagat, A. Kuria, and A. Diallo, "Marker assisted introgression of opaque2 gene into herbicide resistant elite maize inbred lines," African Journal of Biotechnology, vol. 5, no. 24, pp. 2417–2422, 2006.

264. E. Khavkin and E. Coe, "Mapped genomic locations for developmental functions and QTLs reflect concerted groups in maize (Zea mays L.)," Theoretical and Applied Genetics, vol. 95, no. 3, pp. 343–352, 1997.

265. J. L. Araus, G. A. Slafer, C. Royo, and M. D. Serret, "Breeding for yield potential and stress adaptation in cereals," Critical Reviews in Plant Sciences, vol. 27, no. 6, pp. 377–412, 2008.

266. L. R. Veldboom and M. Lee, "Genetic mapping of quantitative trait loci in maize in stress and nonstress environments. II. Plant height and flowering," Crop Science, vol. 36, no. 5, pp. 1320–1327, 1996.

267. J.-M. Ribaut, D. A. Hoisington, J. A. Deutsch, C. Jiang, and D. Gonzalez-de-Leon, "Identification of quantitative trait loci under drought conditions in tropical maize. 1. Flowering parameters and the anthesis-silking interval," Theoretical and Applied Genetics, vol. 92, no. 7, pp. 905–914, 1996.

268. M. Sari-Gorla, M. Villa, C. Frova, P. Krajewski, and N. Di Fonzo, "Genetic analysis of drought tolerance in maize by molecular markers. II. Plant height and flowering," Theoretical and Applied Genetics, vol. 99, no. 1-2, pp. 289–295, 1999.

269. L. Moreau, A. Charcosset, and A. Gallais, "Experimental evaluation of several cycles of marker-assisted selection in maize," Euphytica, vol. 137, no. 1, pp. 111–118, 2004.

270. M. Reymond, B. Muller, A. Leonardi, A. Charcosset, and F. Tardieu, "Combining quantitative trait loci analysis and an ecophysiological model to analyze the genetic variability of the responses of maize leaf growth to temperature and water deficit," Plant Physiology, vol. 131, no. 2, pp. 664–675, 2003.

271. W. D. Beavis, "The power and deceit of QTL experiments: lessons from comparative QTL studies," inProceedings of the 49th Annual Corn and Sorghum Industry Research Conference, pp. 250–266, ASTA, Chicago, Ill, USA, 1994.

272. J. L. Araus, J. Bort, P. Steduto, D. Villegas, and C. Royo, "Breeding cereals for mediterranean conditions: ecophysiological clues for biotechnology application," Annals of Applied Biology, vol. 142, no. 2, pp. 129–141, 2003.

273. M. E. Sorrells, "Application of new knowledge, technologies, and strategies to wheat improvement,"Euphytica, vol. 157, no. 3, pp. 299–306, 2007.

274. R. Tuberosa, S. Salvi, S. Giuliani, et al., "Genome-wide approaches to investigate and improve maize response to drought," Crop Science, vol. 47, pp. S120–S141, 2007.

275. M. Maccaferri, M. C. Sanguineti, S. Corneti, et al., "Quantitative trait loci for grain yield and adaptation of durum wheat (Triticum durum Desf.) across a wide range of water availability," Genetics, vol. 178, no. 1, pp. 489–511, 2008.

276. J. Bernier, A. Kumar, V. Ramaiah, D. Spaner, and G. Atlin, "A large-effect QTL for grain yield under reproductive-stage drought stress in upland rice," Crop Science, vol. 47, no. 2, pp. 507–518, 2007.

277. R. Kumar, R. Venuprasad, and G. N. Atlin, "Genetic analysis of rainfed lowland rice drought tolerance under naturally-occurring stress in eastern India: heritability and QTL effects," Field Crops Research, vol. 103, no. 1, pp. 42–52, 2007.

278. Y. Xu, R. Babu, D. J. Skinner, B. S. Vivek, and J. H. Crouch, "Maize mutant opaque2 and the improvement of protein quality through conventional and molecular approaches," in Proceedings of

International Symposium on Induced Mutations in Plants, IAEA, Vienna, Austria, August 2008.

279. M. A. Lopes, K. Takasaki, D. E. Bostwick, T. Helentjaris, and B. A. Larkins, "Identification of two opaque2 modifier loci in quality protein maize," Molecular and General Genetics, vol. 247, no. 5, pp. 603–613, 1995.
280. D. R. Holding, B. G. Hunter, T. Chung, et al., "Genetic analysis of opaque2 modifier loci in quality protein maize," Theoretical and Applied Genetics, vol. 117, no. 2, pp. 157–170, 2008.
281. W. Yang, Y. Zheng, W. Zheng, and R. Feng, "Molecular genetic mapping of a high-lysine mutant gene (opaque-16) and the double recessive effect with opaque-2 in maize," Molecular Breeding, vol. 15, no. 3, pp. 257–269, 2005.
282. P. Zheng, W. B. Allen, K. Roesler, et al., "A phenylalanine in DGAT is a key determinant of oil content and composition in maize," Nature Genetics, vol. 40, no. 3, pp. 367–372, 2008.
283. G. G. Yousef and J. A. Juvik, "Comparison of phenotypic and marker-assisted selection for quantitative traits in sweet corn," Crop Science, vol. 41, no. 3, pp. 645–655, 2001.
284. S. Gao, C. Martinez, D. Skinner, A. Krivanek, J. H. Crouch, and Y. Xu, "Development of a single-seed based genotyping system for marker assisted selection in maize," Molecular Breeding, vol. 22, pp. 477–494, 2008.
285. Y. Xu, R. Babu, Z. Hao, et al., "SNP-chip based genomewide scan for germplasm evaluation and marker-trait association integrated with selective genotyping and pooled DNA analysis," in Maize Genetics Conference, vol. 51, p. T35, 2009.
286. S. J. Emrich, S. Aluru, Y. Fu, et al., "A strategy for assembling the maize (Zea mays L.) genome,"Bioinformatics, vol. 20, no. 2, pp. 140–147, 2004.
287. C. R. Shyu, J. Green, C. Farmer, et al., "A framework of complex queries for plant phenotype image databases," in The International Conference on Plant, Animal, and Microbe Genomes, p. 875, 2006.
288. S. L. Dwivedi, J. H. Crouch, D. J. Mackill, et al., "The molecularization of public sector crop breeding: progress,

problems, and prospects," Advances in Agronomy, vol. 95, pp. 163–318, 2007.

289. B. A. Kronmiller and R. P. Wise, "TEnest: automated chronological annotation and visualization of nested plant transposable elements," Plant Physiology, vol. 146, no. 1, pp. 45–59, 2008.

290. S. Schroeder, H. Sanchez-Villeda, S. Flint-Garcia, et al., "Integrated software for SNP discovery in maize," inProceedings of the Maize Genetics Conference, vol. 48, p. 50, 2006.

291. R. Tuberosa, S. Salvi, M. C. Sanguineti, P. Landi, M. Maccaferri, and S. Conti, "Mapping QTLS regulating morpho-physiological traits and yield: case studies, shortcomings and perspectives in drought-stressed maize," Annals of Botany, vol. 89, pp. 941–963, 2002.

292. Y. Xu and J. H. Crouch, "Genomics of tropical maize, a stable food and feed across the world," in Genomics of Tropical Crop Plants, P. H. Moore and R. Ming, Eds., pp. 333–370, Springer, London, UK, 2008.

293. R. Ortiz, J. H. Crouch, M. Iwanaga, et al., "Agriculture and energy in developing countries: bio-energy and Agricultural Research-for-Development," IFPRI "2020 Focus" Policy Brief #14, 2006,http://www.ifpri.org/pubs/catalog.htm#focus.

Chapter 2

Genetic Enhancement of Grain Quality-Related Traits in Maize

H. Hartings[1], M. Fracassetti[1] and M. Motto[1]

[1]CRA-Unità di Ricerca per la Maiscoltura, Bergamo, Italy

INTRODUCTION

Maize (*Zea mays*) is a major food and animal feed worldwide and occupies a relevant place in the world economy and trade as an industrial grain crop (White & Johnson, 2003). Currently more than 70% of maize production is used for food and feed; therefore, knowledge of genes involved in grain structure and chemical is important for improving the nutritional and food-making properties of maize.

Although, plant breeding has been extremely successful at improving the yield of maize, quality has received less attention. However, important advances were made by breeders in this area as well, resulting in maize with a wide range of compositions. In fact, by exploiting genetic variation, the composition of the kernel was altered for both the quantity and quality (structure and chemical diversity) of starch, protein, and oil throughout kernel development. Furthermore, the ability of plant scientists to use existing genetic variation and to identify and manipulate commercially important genes will open new avenues to design novel variation in grain composition. This will provide the basis for the development of the next generation of speciality in maize and of new products to meet future needs.

This chapter focuses on gene discovery, exploitation, and genetic variation known to affect the development and chemical composition of maize kernel. Throughout the chapter we have attempted to summarize the current status in these areas with a particular reference to deposition of storage proteins, starches, lipids, and carotenoids, and research pertinent to enhance kernel quality-related traits. Finally, we provide a brief outlook on future developments in this field and the resultant opportunities and application of conventional and molecular breeding for the development of new maize products better suited to its various end uses.

Kernel Growth and Development

The great economical and nutritional value of the maize kernel is mainly due to its high starch content, as it represents approximately 75% of the mature seed weight. However, the protein complement (ca. 10% of the mature seed weight), mainly found in the form of zeins (storage proteins) is essential for human and animal nutrition. Yet the question remains of why the selection for higher starch level irremediably results in less protein content, as illustrated by the Illinois Long-Term Selection Experiment, which is spanning over more than 100 generations of classical breeding (Mooses et al., 2004).

As a typical angiosperm, the maize kernel comprise two zygotic tissues, namely the embryo (germ) and the endosperm, that are embedded in the testa (or seed coat) and the pericarp (or fruit wall), which fuse into a thin protective envelope. The endosperm is the main storage site of starches and proteins, whereas the embryo reserves mainly lipids. However, the economical and nutritional value of the kernel is mostly derived from the endosperm, a starch-rich tissue, that supports the embryo at germination.

In maize, endosperm makes up the majority of kernel dry matter (70-90%) and is the predominant sink of photosynthates and other assimilates during reproductive growth; therefore, factors that mediate endosperm development to a large extent also determine grain yield. Furthermore, the endosperm of seed can serve as a valuable system to address fundamental questions related to the improvement of seed size in crops.

High-throughput genomics and post-genomics approaches are now providing new tools for a better understanding of the genetic and biochemical networks operating during kernel development. Recently, large databases of maize gene expressed sequence tags (ESTs) have been made available (i.e http://www.maizegdb.org), and transcriptome analyses aimed at identifying genes involved in endosperm development and metabolism have been published, along with computer software to systematically characterize them, has made possible to analyze gene expression in developing maize endosperm more thoroughly to identify tissue-specific genes involved in endosperm development and metabolism (Lai et al., 2004; Verza et al., 2005; Liu et al., 2008; Prioul et al., 2008). These studies have shown that in maize, at least 5000 different genes could be expressed during development. However, about 35% of them are orphan genes, whose functions remain enigmatic, possibly corresponding to endosperm specific genes (Liu et al., 2008), as also observed in wheat (Wan et al., 2008). Furthermore, Mechin and co-workers (2004) have established a proteome reference map for the maize endosperm. They found that metabolic processes, protein destination and synthesis, cell rescue, defence, cell death and ageing were the most abundant functional categories detected in the maize endosperm.

Collectively, the transcriptome and proteome maps constitute a powerful tool for physiological studies and are the first step for investigating maize endosperm development and metabolism. Although, mRNAs are the primary products of gene expression and their levels are often weakly correlated to corresponding protein levels (Gygi et al., 1999), the analyses of the changes in the transcription profiles of endosperm mutants may allow to formulate predictions regarding the biological role of these loci in endosperm development and metabolism. This information is useful for identifying distinctive, previously uncharacterized, endosperm-specific genes; in addition, it provides both further research material for academic laboratories, and material for plant breeders and food processors to include in their respective research or product pipelines.

Accumulation of Storage Products

The structure and biochemical properties of seed storage compounds have been widely investigated over the past 30 years due to their abundance, complexity, and impact on the overall nutritional value of the maize seed. A great deal is now known about the compounds that are made and stored in seeds, as well as how they are hydrolyzed and absorbed by the embryo. For more detailed reviews describing the nature and biochemistry of maize endosperm and embryo storage products, we refer the reader to a number of recent reviews (i.e. Hannah, 2007; Holding & Larkins, 2009; Motto et al., 2009; Val et al. 2009).

STORAGE PROTEIN

The primary storage proteins in the maize grain are prolamines called "zeins". Specifically, the zeins are the most abundant protein storage component (>60%) in developing endosperm tissues and are constituted by alcohol-soluble compounds with a characteristic amino acid composition, being rich in glutamine, proline, alanine, and leucine, and almost completely devoid of lysine and tryptophan (Gibbon & Larkins, 2005). From a nutritional point of view, the exceedingly large proportion of codons for hydrophobic amino acids in α-zeins is mostly responsible for the imbalance of maize protein reserves. Therefore, the reduction in α-zein protein accumulation with biased amino acid content could provide a correction to this imbalance. Zeins have also unique functional and biochemical properties that make them suitable for a variety of food, pharmaceutical, and manufactured goods (Lawton, 2002). Based on their evolutionary relationships, zeins are divided into four protein subfamily of α- (19 and 22-kDa), β-

(15 kDa), γ- (16-, 27-, and 50-kDa), and δ-zeins (10- and 18-kDa), that are encoded by distinct classes of structural genes (Holding & Larkins, 2009). Miclaus et al. (2011) have recently reported that -zein genes have evolved from a common ancestral copy, located on the short arm of chromosome 1, to become a 41-member gene family in the reference maize genome, B73. According to these workers once genes are copied, expression of donor genes is reduced relative to new copies. In particular, epigenetic processes that modify the information content of the genome without changing the DNA sequence, seems to contribute to silencing older copies: some of them can be reactivated when endosperm is maintained as cultured cells, indicating that copy number variation might contribute to a reserve of gene copies.

The proper deposition of zeins inside subcellular structures called protein bodies (PBs) confers the normal vitreous phenotype to the endosperm. PBs are specialized endosperm organelles that form as an extension of the membrane of the rough endoplasmic reticulum (RER), into which zeins are secreted as the signal peptide is processed. After being secreted into the RER, the β- and γ-zeins form a matrix, which is penetrated by the α- and δ-zeins, enlarging the PB and making it a spherical structure of 1-2 μm (Lending & Larkins 1989). Alterations in size, shape or number of PBs generally determine the opaque phenotype (Holding and Larkins 2009), the sole exception being *floury1 (fl1)*, an opaque mutant with no alterations in PB size or shape (Holding et al. 2008). Recently, maize storage protein mutants created through RNAi showed that γ- zein RNA interference (RNAi) maize mutant lines exhibited slightly altered PB body formation and that a more drastic effect was observed in the β-γ - combined mutant, where protein bodies showed an irregular shape, particularly in their periphery (Wu & Messing, 2010). Further studies reported by Llop-Touset et al., (2010) have indicated that the N-terminal proline-rich domain of γ-zein plays an important role in PB formation. To gain a deeper insight into the relationship between RNA and protein localization in plants, Washida et al. (2009) have identified that the cis-localization elements of the 10-kDa δ-zein are responsible for PB-ER targeting. Their results indicate that there is a close relationship between RNA and protein localization in plant cells and that RNA localization may be an important process in mediating the deposition of storage protein in the endomembrane system in plants.

Endosperm Mutants Altering Storage Protein Synthesis

As highlighted before, endosperm growth and development is a complex phenomenon that may be driven by the coordinate expression of numerous genes. Strategies using spontaneous and induced mutants allow the characterization of the complex underlying gene expression system integrating carbohydrate, amino acid, and storage protein metabolisms and operating during endosperm growth and development. In this respect several endosperm mutants altering the timing and the rate of zein synthesis have been described (reviewed by Motto et al., 2009). The mutants altering the rate of zein synthesis exhibit a more or less defective endosperm and have a lower than normal zein content at maturity. Many of these genes have been mapped to chromosomes and their effect on zein synthesis has been described (Table 1). All mutants confer an opaque phenotype to the endosperm, and, as zein synthesis is reduced, the overall lysine content is elevated, giving potential for use in the development of «high-lysine» maize.

Table 1: Some features of maize mutans affecting zein accumulation

Genotype	Inheritance	Effect on zein accumulation	Molecular bases
Opaque-2 (o2)	Recessive	22-kDa elimination, 20-kDa reduction,	Transcriptional activator
Opaque-5 (o5)	Recessive	No reduction	MGD1
Opaque-6 (o6)	Recessive	General reduction	
Opaque-7 (o7)	Recessive	General reduction 20 and 22-kDa	ACS-like protein
Opaque-15 (o15)	Recessive	27-kDa reduction, reduction -zein	
Opaque-2 modifiers	Semidominant	27-kDa overproduction	
Floury-1 (fl1)	Semidominant	General reduction	Transmembrane protein

Floury-2 (fl2)	Semidominant	General reduction	Defect 22-kDa zein
Floury-3 (fl3)	Semidominant	General reduction	
Defective-endo-spermB30 (De*B30)	Dominant	General reduction	Defect 20-kDa zein
Mucronate (Mc1)	Dominant	General reduction	Abnormal 16-kDa -zein
Zpr10(22)	Recessive	10-kDa reduction	

Genetics has played an important role in discovering a series of opaque endosperm mutants and demonstrating their effects on genes mediating zein deposition (Motto et al., 2009). For example, the recessive mutation *opaque-2* (*o2*) induce a specific decrease in the accumulation of 22-kDa α-zeins, while the *opaque-15* (*o15*) mutation exerts its effect primarily on the 27-kDa γ-zeins. The *floury1* (*fl1*) mutation is somewhat different, since it does not affect the amount or composition of zein proteins but rather results in the abnormal placement of α-zeins within the PB: *Fl1* encodes a transmembrane protein that is located in the protein body ER membrane. Similarly, Myers et al. (2011) have found that the *opaque5* (*o5*) mutant phenotype is caused by a reduction in the galactolipid content of the maize endosperm, with no change in zein proteins. Furthermore, these workers reported that *O5* locus encodes the *monogalactosyldiacylglycerol synthase (MGD1)* and specifically affects galactolipids necessary for amyloplast and chloroplast function. A further interesting maize opaque endosperm mutant, termed *mto140*, which also shows retarded vegetative growth has been studied by Holding et al., (2010). The seeds showed a general reduction in zein storage protein accumulation and an elevated lysine phenotype typical of other opaque endosperm mutants; however, it is distinct from the other opaque mutants because it does not result from quantitative or qualitative defects in the accumulation of specific zeins but rather from a disruption in amino acid biosynthesis. Because the opaque phenotype co-segregated with a Mutator transposon insertion in an *arogenate dehydrogenase* gene (*zmAroDH-1*), this has led the previous authors the characterization of the four-member family of maize *arogenate dehydrogenase* genes (*zmAroDH-1–4*) which share highly similar sequences. Their differential expression patterns, as well as subtle mutant effects on the accumulation of tyrosine and phenylalanine in endosperm, embryo, and leaf tissues, suggested that the functional redundancy of this gene family provides metabolic plasticity for the synthesis of these important amino acids.

The *o2* mutation has been widely studied at the genetic, biochemical and molecular levels. *O2* encodes a basic leucine zipper (*bZIP*) transcriptional regulator that is specifically expressed in the endosperm (reviewed in Motto et al., 2009). These studies showed that *O2* activates the expression of 22-kDa α-zein and 15-kDa β-zein genes by interacting with the TC-CACGT(a/c)R(a/t) and GATGYRRTGG sequences of their promoters, therefore displaying a broad binding specificity and recognizing a variety of target sites in several distinct genes. *O2* also regulates directly or indirectly a number of other non-storage protein genes, including *b-32*, encoding a type I ribosome-inactivating protein, one of the two cytosolic isoforms of the *pyruvate orthophosphate dikinase* gene (*cyPPDK1*), and *b-70*, encoding a heat shock protein 70 analogue, possibly acting as a chaperonin during PB formation. *O2* also regulates the levels of *lysine-ketoglutarate reductase* (Brochetto-Braga et al., 1992) and *aspartate kinase1* (Azevedo et al., 1997). These broad effects suggest that *O2* plays an important role in the developing grain as a coordinator of the expression of genes controlling storage protein, and nitrogen (N) and carbon (C) metabolism.

The *O7* gene was recently cloned by two different groups, using a combination of map-based cloning and transposon tagging and confirmed by transgenic functional complementation (Miclaus et al., 2011;Wang et al., 2011). Moreover, these last workers via sequence analysis indicated that the *O7* gene showed similarities with members of the larger family of *acyl-CoA synthetase*-like genes (*ACS*), although its exact enzymatic activity is uncertain. In particular, Miclaus et al (2011), have hypothesized a mechanism in which the O7 protein functions in post-translational modification of zein proteins, thus contributing to membrane biogenesis and stability of PBs and conferring the normal vitreous phenotype of the kernel. Alternatively, Wang et al. (2011) have suggested, by analysis of amino acids and key metabolites, that *O7* gene function might affect amino acid biosynthesis by affecting α-ketoglutaric acid and oxaloacetic acid phenotype, indicating a conserved biological function of *O7* in cereal crops. In this respect, Hartings et al. (2011), in a study to clarify the role that *O2* and*O7* play in endosperm gene expression through transcriptomic analyses, indicated that the *o2* and *o7*mutants alter gene expression in a number of enzymatic steps in the tricarboxylic acid cycle (TCA) and glycolysis pathways that are of central importance for the amino acid metabolism in developing seeds. Although, a systematic characterization of such enzymes will be necessary before any inferences are warranted, the cloning of *O7* revealed a novel regulatory mechanism for storage protein synthesis and highlighted an effective target for the genetic manipulation of storage protein contents in cereal seeds, maize included.

An alternative approach to understand the relationship between zein synthesis and the origin of the opaque endosperm phenotype is to perturb zein accumulation transgenically. In this respect, a number of laboratories have reported a reduction in 22-kDa (Segal et al., 2003) and 19-kDa α-zeins (Huang et al., 2004) by RNAi and by seed-specific expression of lysine rich protein (Rascon-Cruz et al., 2004;Yu et al., 2004).

Regulation of Storage Protein Synthesis

The expression of zein genes is regulated coordinately and zein mRNAs accumulate at high concentrations during early stages of endosperm development (reviewed in Motto et al., 2009). From these studies it was also noted that the coordinate expression of zein genes in maize is controlled primarily at the level of transcription according to specific spatial/temporal patterns. Therefore, attention has turned to understanding the regulatory mechanisms responsible for zein gene expression. Highly conserved *cis*-regulatory sequences have been identified in the promoter of prolamine genes and corresponding *trans*-activity factors (cf Motto et al., 2009). Zein gene expression can also be affected by other regulatory mechanisms, such as methylation, amino acid supply and phosphorylation. In this context, Locatelli et al. (2009) have provided evidence that O2-mediated transcriptional activation occurs in two-phases, first a potentiated and second a transcriptional activated phase, both characterized by a specific profile of chromatin modifications. The dependency on O2 activity in the establishment of these chromatin states was different for distinct sub-sets of *O2* targets, indicating a gene-specific interaction of *O2* with chromatin modifying mechanisms in driving transcription.

Practical Applications and Perspectives

Despite efforts to develop opaque mutations that are commercially useful, its inherent phenotypic deficiencies, such as soft endosperm texture, lower yield, increased seed susceptibility to pathogens and mechanical damages, have limited their use. To overcome these drawbacks Quality Protein Maize (QPM) strains were created by selecting genetic modifiers that convert the starchy endosperm of an o2 mutant to a hard, vitreous phenotype. Genetic studies have shown that there are multiple, unlinked *o2*modifiers (*Opm*), (review in Gibbon and Larkins, 2005). Genetic analysis of *o2* modifiers identified several disperse quantitative trait loci (QTLs). Although their molecular identities have remained unknown, QTLs could be correlated with observed increases in 27-kDa γ-zein transcript and protein in QPM (Holding et al. 2008, and references therein). Two different QTLs, which are candidates for *o2*modifier genes, affect 27-kDa γ-zein gene expression. The first of these is associated with increased expression and the other is linked to *o15*, a mutation at a different chromosome 7 location, which causes decreased 27-kDa γ -zein expression suggesting that the amount of γ-zeins would become critical to keep starch granules embedded in the vitreous area. To examine the role of γ-zeins in QPM, Wu et al., (2010) have used an RNAi construct, designed from the inverted coding sequences of the 27-kDa γ-zein gene, to knock down both 27- and 16-kDa γ-zeins by taking advantage of their DNA sequence conservation. Their findings reinforce the fact that different zeins have evolved to play distinct roles in the development of the endosperm.

Although maize endosperm storage protein genes have been studied for many years, many questions regarding their sequence relationships and expression levels have not been solved, such as structure, synthesis and assembly into protein bodies, and their genetic regulation (Holding and Larkins, 2009). The development of tools for genome-wide studies of gene families makes a comprehensive analysis of storage protein gene expression in maize endosperm possible with the identification of novel seed proteins that were not described previously (Woo et al., 2001). For example, to advance our understanding of the nature of the mutations associated with an opaque phenotype, Hunter et al. (2002) assayed the patterns of gene expression in a series of opaque endosperm mutants by profiling endosperm mRNA transcripts with an Affimetrix GeneChip containing approximately 1,400 selected maize gene sequences. Their results revealed distinct, as well as shared, gene expression patterns in these mutants. Similar research on the pattern of gene expression in *o2*, *o7*, and in the *o2o7* endosperm mutants was carried out by Hartings et al. (2011) by profiling endosperm mRNA transcripts at 14 DAP. Their results, based on a unigene set composed of 7,250 ESTs, allowed to identify a series of mutant related up-regulated (17.1%) and down-regulated (3.2%) transcripts. In addition, the same authors identified several differentially expressed ESTs, homologous to gene encoding enzymes involved in amino acid synthesis, C metabolism (TCA cycle and glycolysis), storage protein and starch metabolism, gene transcription and translation processes, signal transduction, and in protein, fatty acid, and lipid synthesis. Those analyses demonstrate that the mutants investigated are pleiotropic and play a critical role in several endosperm metabolic processes. Although, by necessity, these data are descriptive and more work is required to define gene functions and dissect the complex regulation of gene expression, the genes isolated and characterized to date give us an intriguing insight into the mechanisms underlying amino acid metabolim in the endosperm.

A useful strategy to develop more quickly new QPM varieties has been proposed by Wu and Messing (2011). In fact, conversion of QPM into local germplasm is a lengthy process that discourages the spread of the benefits of QPM because breeders have to monitor a high-lysine level, the recessive *o2*mutant allele, and the modifiers *o2*, (M*o2*s). Accordingly, to overcome this problem these last authors presented a simpler and accelerated QPM selection. Instead of using the recessive *o2* mutation, they used an RNAi construct directed against both 22- and 19-kDa zeins, but linked to the visible green fluorescent protein (GFP) marker gene. Indeed, when such a green and nonvitreous phenotype was crossed with QPM lines, the Mo2s produced a vitreous green kernel, demonstrating that high lysine and kernel hardness can be selected in a dominant fashion.

Starch Synthesis

Maize, like other cereals, accumulate starch in the seed endosperm as an energy reserve. Moreover, its starch is one of the most important plant products and has various direct and indirect applications in food, feed, and industries. For this reason attempts to increase starch accumulation have received a great deal of attention by plant breeders and plant scientists. Starch biosynthesis is a central function in plant metabolism that is accomplished by a multiplicity of conserved enzymatic activities (see Hannah & James 2008, for a review). Roughly three-quarters of the total starch is amylopectin, which consists of branched glucose chains that form insoluble, semi-crystalline granules. The remainder of the starch is amylose, which is composed of linear chains of glucose that adopt a helical configuration within the granule (Myers et al., 2000). Briefly starch synthesis has two fundamental activities represented by starch synthase, which catalyzes the polymerization of glucosyl units into α (1/4)-linked "linear" chains, and starch-branching enzyme, which catalyzes the formation of α (1/6)-glycoside bond branches that join linear chains. Acting together, the starch synthases and starch-branching enzymes assemble the relatively highly branched polymer amylopectin, with approximately 5% of the glucosyl residues participating in α (1/6)-bonds, and the lightly branched molecule amylose. A third activity necessary for normal starch biosynthesis is provided by starch-debranching enzyme (DBE), which hydrolyzes α (1/6)-linkages. Two DBE classes have been conserved separately in plants (Beatty et al., 1999). These are referred as pullulanase-type DBE (PUL) and isoamylase-type DBE (ISA), based on similarity to prokaryotic enzymes with particular substrate specificity. ISA functions in starch production are implied from genetic observations that mutations typically result in reduced starch content, abnormal amylopectin structure, altered granule morphology, and accumulation of abnormally highly branched polysaccharides similar to glycogen.

Genes Affecting Starch Biosynthesis

Starch biosynthesis in seeds is dependent upon several environmental, physiological, and genetic factors (reviewed in Boyer and Hannah, 2001). Moreover, the maize kernel is a suitable system for studying the genetic control of starch biosynthesis. A large number of mutations that cause defects in various steps in the pathway of starch biosynthesis in the kernel have been described. Their analysis has contributed greatly to the understanding of starch synthesis (reviewed in Boyer and Hannah, 2001). In addition, these mutations have facilitated the identification of many genes involved in starch biosynthetic production.

As there seems little point in reviewing these data, we will simply summarize in Table 2 cloned maize genes and their gross phenotypes. Although, the effects shown in this table may not necessarily be the primary effect of a mutant, these are the ones presently known. More recently, Kubo et al. (2010) have described novel mutations of *sugary1 (su1)* and *isa2* loci, coding for isoamylase-type starch-DE enzyme (ISA) ISA1 and ISA2, respectively: Their data indicate that in maize endosperm these enzymes function to support starch synthesis either as a heteromeric multisubunit complex containing both ISA1 and the noncatalytic protein ISA2 or as a homomeric complex containing only ISA1. In particular, it was found that i) homomeric ISA has specific functions that determine amylopectin structure that are not provided by heteromeric ISA and ii) tissue-specific changes in relative levels of ISA1 and ISA2 transcripts, or functional changes in the ISA1 protein, could explain how maize endosperm acquired the homomeric enzyme.

Table 2: Summary of mutant effects in maize where an associated enzyme lesion has been reported.[a] Changes relative to normal. ↑, ↓ = increase or decrease, respectively. Sugars = the alcohol-soluble sugars

Genotype	Mayor biochemical changesa	Enzyme affected
Shrunken-1 (sh1)	↑ Sugars, ↓ Starch	↓ Sucrose synthase
Shrunken-2 (sh2)	↑ Sugars, ↓ Starch	↓ ADPG-pyrophosphorylase, ↑ Hexokinase
Brittle-1 (bt1)	↑ Sugars, ↓ Starch	↓ Starch granule-bound phospho-oligosaccharide synthase
Brittle-2 (bt2)	↑ Sugars, ↓ Starch	↓ ADPG-pyrophosphorylase
Shrunken-4 (sh4)	↑ Sugars, ↓ Starch	↓ Pyridoxal phosphate
Sugary-1 (su)	↑ Sugars, ↓ Starch	↑ Phytoglycogen branching enzyme, ↓ Phytoglycogen debranching enzyme
Waxy (wx)	↑ 100% Amylopectin	↓ Starch-bound starch syntase, ↑ Phytoglycogen branching enzyme
Amylose-extender(ae)	↑ Apparent amylose, ↑ Loosely branched polysaccharide	↓ Branching enzyme IIb

Dull-1 (du1)	↑ Apparent amylase	↓ Starch synthase II, ↓ Branching enzyme Iia,↑ Phytoglycogen branching enzyme

Many biochemical and molecular studies on starch synthesis have been also focused on identifying the rate limiting enzymes to control metabolism. In this context, ADP-glucose pyrophosphorylase (AGPase) plays a key role in regulating starch biosynthesis in cereal seeds. The AGPase in the maize endosperm is a heterotetramer of two small subunits encoded by *Brittle2* (*Bt2*) gene, and two large subunits, encoded by the *Shrunken2* (*Sh2*) gene. Transgenic approaches focused on allosteric regulation of AGPase, although studies of the kinetic mechanism of maize endosperm AGPase has uncovered complex regulatory properties (Kubo et al., 2010), increase starch content and caused an increased seed weight than lines expressing wild-types (Giroux et al., 1996; Wang et al., 2007). Additional research has been also devoted to the over-expression of the wide-type genes encoding maize AGPase. For example Li et al. (2011), have transferred the *Bt2* and *Sh2* genes from maize, with an endosperm-specific promoter from 27-kDa zein or an endosperm-specific promoter from 22-kDa zein, into elite inbred lines, solely and in tandem, by *Agrobacterium tumefaciens*-mediated transformation. They found that developing transgenic maize kernels exhibited higher *Bt2* and *Sh2* gene expression, higher AGPase activity, higher seed weight, and the kernels accumulated more starch compared with non-transgenic plants. The over-expression of either gene enhanced AGPase activity, seed weight (+15%) and starch content compared with the wild type, but the amounts were lower than plants with over-expression of both *Bt2* and *Sh2*. Collectively, these results indicate that over-expression of those genes in transgenic maize plants could improve kernel traits and provide a feasible approach for enhancing starch content and seed weight in maize.

Regulation of Starch Synthesis

In spite of the above mentioned studies as a complex metabolic pathway, the regulation of starch biosynthesis is still poorly understood. This is surprising, considering the number and variety of starch mutations identified so far, which may indicate that nutrient flow is the key regulatory stimulus in carbohydrate interconversion. In this connection, it has been argued that glucose also serves as a signal molecule in regulating gene expression, in some cases, different sugars or sugar metabolites might act as the actual signal molecules (reviewed in Koch, 2004). There is evidence that regulation of major grain-filling pathway is highly integrated in endosperm. Gene responses to sugars and C/N balance have been implicated. For example, Sousa et al. (2008) have recently identified in maize a gene for*Sorbitol dehydrogenase1 (Sdh1)*. They showed that this gene is highly expressed early in seed development throughout the endosperm, with greatest levels in the basal region, compatible with SDH involvement in the initial metabolic steps of carbohydrate metabolism. The same authors also presented genetic, kinetic, and transient expression evidence for regulation at the transcriptional level by sugars and hypoxia. Moreover, many pleiotropic defective kernel (*dek*) mutations that fail to initiative or complete grain-filling have been identified, but not studied in detail. These are likely to include mutations in "housekeeping genes" as well as important developmental mutants or transcription factors. In this respect, a key challenge is to devise molecular and genetic strategies that can be used to effectively analyse this large, complex phenotypic class. As far as transcription factors are concerned,Fu & Xue (2010) have recently identified in rice candidate regulators for starch biosynthesis by gene coexpression analysis. Among these genes, *Rice Starch Regulator1 (RSR1)*, an APETALA2/ethylene-responsive element binding protein family transcription factor, was found to negatively regulate the expression of type I starch synthesis genes; moreover, *RSR1* deficiency results in the enhanced expression of starch synthesis genes in seeds. Collectively these results demonstrate the potential of co-expression analysis for studying rice starch biosynthesis and the regulation of a complex metabolic pathway and provide informative clues, including the characterization of *RSR1*, to facilitate the improvement of seed quality and nutrition. It is expected that similar orthologous loci will be soon identified in maize; this will allow us to deeper our knowledge on regulatory mechanisms affecting starch biosynthesis in maize.

Different approaches in this area are needed to identify direct interaction among starch biosynthetic enzymes, as well as modifying factors that regulate enzyme activity. In this respect, Wang et al. (2007) described a study in which a bacterial *glgC16* gene, which encodes a catalytically active allosteric-insensitive enzyme, was introduced into maize. The results of this study showed that developing transgenic maize seeds exhibited higher AGPase activity (a rate limiting step in glycogenesis and starch synthesis), in the presence of an inhibitory level of Pi in vitro, compared with the untransformed control. More interestingly, the same authors found the seed weight of transgenic plants was increased significantly. Furthermore, tools for genome-based analyses of starch biosynthesis pathway are now available for maize and other cereals. This may eventually help to explain species differences in starch granule shape and size, and thus provide the potential for agricultural advances. Recently, Prioul et al. (2008) have provided information on carbohydrate metabolism by comparing gene expression at three levels - transcripts, proteins and enzyme activities - in relation to substrate or product in developing kernels from 10 to 40 DAP. Their study have identified two distinct patterns: during endosperm development: invertases and hexoses are predominant at the beginning, whereas enzyme patterns in the starch pathway, at the three levels, anticipate and parallel starch accumulation, suggesting that, in most cases, transcriptional control is responsible for the regulation of starch biosynthesis.

LIPIDS

While intensive agricultural and industrial uses of the maize kernel is widely due to its high starch content, the oil stored in the maize kernel also has considerable importance. Moreover, its oil is the most valuable co-product from industrial processing of maize grain through wet milling or dry milling and is high-quality oil for human.

Research in this field (for review see Val et al., 2009) indicate that i) the mature embryo is approximately 33% lipid in standard hybrids and contains about 80% of the kernel lipids; ii) high-oil maize shows a greater feed efficiency than normal-oil maize in animal feed trials: the caloric content of oil is 2.25 times greater than that of starch on a weight basis and its fatty acid composition, mainly oleic and linoleic acids; iii) maize oil is highly regarded for its low level of saturated fatty acids, on average 11% palmitic acid and 2% stearic acid, and its relatively high levels of polyunsaturated fatty acids such as linoleic acid (24%); and iv) maize oil is relatively stable, since it contains only small amounts of linolenic acid (0.7%) and high levels of natural antioxidants. Additionally, it was found that oil and starch are accumulated in different compartments of the maize kernel: 85% of the oil is stored in the embryo, whereas 98% of the starch is located in the endosperm. Therefore, the relative amounts of oil and starch are correlated with the relative sizes of the embryo and endosperm and successful breeding for high oil content in the Illinois High Oil strains has mainly been achieved through an increase in embryo size (Moose et al., 2004). Whereas the embryo represents less than 10% of the kernel weight in normal or high-protein lines, it can contribute more than 20% in high-oil lines. However, genetic components may also modulate oil content in the embryo, independently of its size, as shown by the cloning of a high-oil QTL in maize that is caused by an amino acid insertion in an acyl-CoA:diacylglycerol acyltransferase catalyzing the last step of oil biosynthesis (Zheng et al., 2008).

Lipid Biosynthetic Pathway and Genetic Inheritance

The primary determinant of amount of lipids in maize kernels is the genetic makeup (Lambert, 2001). In maize studies through genetic mapping of oil traits reported that multiple (>50) QTLs are involved in lipid accumulation (Laurie et al., 2004), making yield improvement through conventional breeding difficulty. High-oil varieties of maize were developed at the University of Illinois through successive cycles of recurrent selection (Dudley and Lambert, 1992). Although these lines have an improved energy content for animal feeding applications, the poor agronomic characteristics, including disease susceptibility and poor standability. These deficiencies precluded their commercial introduction on broad hectarage.

In spite of a good understanding of the oil biosynthetic pathway in plants and of the many genes involved in oil pathway have been isolated, the molecular basis for oil QTL is largely unknown. However, Zheng et al. (2008) have recently found that a oil QTL (*qHO6*) affecting maize seed oil and oleic-acid content, encodes an acyl-CoA:diaglycerol acytransferase (DGAT1-2), which catalyze the final step of oil synthesis.

As far as the composition in concerned, maize oil is mainly composed of palmitic, stearic, oleic, linoleic, and linolenic fatty acids. Evidence has shown that genetic variation existed also for the fatty acid composition of the kernel (Lambert, 2001). In essentially all studies, researchers suggested that major gene effects were being modulated by modifier genes for oil composition. Although it seems that sources of major genes for composition of maize oil can be utilized, other studies indicate that the inheritance of oleic, linoleic, palmitic, and stearic acid content when considered together is complex and under multigenic control (Sun et al., 1978). Molecular characterization of fatty acid desaturase-2 (*fad2*) and fatty acid desaturase-6 (*fad6*) in this plant indicates that *fad2* and *fad6* clones are not associated with QTLs for the ratio of oleic/linoleic acid, suggesting that some of the QTLs for the oleic/linoleic acid ratio do not involved variants of *fad2* and *fad6*, but rather involved other gene that may influence flux via enzymes encoded by *fad2* or *fad6*. Additional studies are needed to more precisely identify the genes and enzymes involved in determining the composition of maize oil. Application of powerful new technologies, such as transcription profiling, metabolic profiling, and flux analyses, should prove valuable to achieving this scope. In addition, identification of transcription factors or other regulatory proteins that exert higher level control of oil biosynthesis or embryo development will be particularly attractive candidate for biotechnology approaches in the future.

In maize, Pouvreau et al. (2011) have recently identified orthologs related, respectively, to the master regulators *LEAFY COTYLEDON1-2 (i.e. ZmLEC1),* that directly activate in Arabidopsis genes involved in TAG metabolism and storage, and to the transcription factor *WRINKLED1* (i.e.. *ZmWRI1a*and *ZmWRI1b),* necessary to mediate the regulatory action of the master regulators towards late glycolytic and oil metabolism. In this crop, both genes are preferentially expressed in the embryo and exhibit a peak of expression at the onset of kernel maturation. *ZmWRI1a* is induced by *ZmLEC1* (Shen et al., 2010). Additionally, transcriptomic analyses carried out on *ZmWRI1a* over-expressing lines have allowed to the previous workers to identify putative target genes of *ZmWRI1a* involved in late glycolysis, fatty acid or oil metabolism. Though not fully overlapping, the sets of *AtWRI1* and*ZmWRI1a* target genes are very resembling. Exhaustive analyses relying on ChIP experiments would allow determining whether these sets are identical. Interestingly, the DNA AW-box proposed to be bound by *AtWRI1* (Maeo et al., 2009) was also identified in promoter sequences of putative target genes of *ZmWRIa*, suggesting that even the *cis*-regulatory element recognized by *WRI1* seems have been conserved between dicots and monocots. Additionally this study has shown that transgenic*ZmWri1a-OE* kernels did not only induce a significant increase in saturated and unsaturated fatty acids with 16 to 18 C atoms but also cause a significant increase for several free amino acids (Lys, Glu, Phe, Ala, Val), intermediates or cofactors of amino acid biosynthesis (pyro-Glu, aminoadipic acid, Orn, nor- Leu), and intermediates of the TCA cycle (citric acid, succinic acid). Since the transcriptome analysis suggests that *ZmWri1a* essentially activates genes coding for enzymes in late glycolysis, fatty acid, CoA, and TAG biosynthesis, and considering that no misregulated candidates participate in any additional pathways, the increase in amino acids and TCA intermediates probably reflects secondary adjustments of the C and N metabolism to the increased oil biosynthesis triggered by *ZmWri1a*. The three amino acids Phe, Ala, and Val are derived from PEP or pyruvate, and their increase may simply be a byproduct of a strongly increased C flux through glycolysis.

Carotenoid Pigments

Along with their essential role in photosynthesis, carotenoids are of significant economic interest as natural pigments and food additives (reviewed in Botella-Pavía & Rodríguez-Concepción, 2006). Their presence in the human diet provides health benefits as nontoxic precursors of vitamin A and antioxidants, including protection against cancer and other chronic diseases (review by Fraser & Bramley 2004). These motives have promoted scientists to explore ways to improve carotenoid content and composition in staple crops (reviewed in Sandmann et al. 2006; Zhu et al. 2009). Analyses of genotypes with yellow to dark orange kernels exhibits considerable natural variation for kernel carotenoids, with some lines accumulating as much as 66 μg/g (e.g. Harjes et al., 2008), with provitamin A activity (β -cryptoxanthin, α - and β -carotene is typically small (15% to 18% of the total carotenoids fraction) compared to lutein or zeaxanthin (45% and 35%, respectively; Kurlich & Juvik, 1999; Brenna and Berardo, 2004). Moreover, a moderate to high heritability estimates indicate that breeding for increased levels of both carotenes and xanthophylls should be feasible.

Carotenoid Biosynthesis and Genetic Control

Carotenoids are derived from the isoprenoid biosynthetic pathway and are precursors of the plant hormone abscisic acid (ABA) and of other apocarotenoids (Matthews and Wurtzel, 2007). In maize characterization of the carotenoid biosynthetic pathway has been facilitated by the analysis of mutants associated with reduced levels of carotenoids. In fact, by using this approach in maize three genes controlling early steps in the carotenoid pathway have been cloned. The use of these cloned genes as probes on mapping populations will enable the candidate gene approach to be used for studying the genetic control of quantitative variation in carotenoids. Accordingly, Wurtzel et al.. (2004) detect major QTLs affecting accumulation of β -carotene and β -cryptoxanthin indicating that these QTLs could be selected to increase levels of pro-vitamin A structures. Chander et al. (2007), using a RIL population found 31 QTL including 23 for individual and 8 for total carotenoid accumulations. Moreover, Harjes et al. (2008), via association mapping, linkage mapping, expression analysis, and mutagenesis, showed that variation in *lycopene epsil cyclase* (*lcyE*) locus alters flux down α -carotene versus β -carotene branches of the carotenoid pathway. Additional experimental evidence obtained byYan et al. (2010) have documented that also the gene encoding β *-carotene hydroxylase1 (crtRB1)*underlies a principal QTL associated with β -carotene concentration and conversion in maize kernels. Moreover, the same workers noted that the *crtRB1* alleles associated with reduced transcript expression correlate with higher β -carotene concentrations. Genetic variation at *crtRB1* also affects hydroxylation efficiency among encoded allozymes, as observed by resultant carotenoid profiles in recombinant expression assays. Similarly, studies on natural maize genetic diversity carried out by Vallabhaneni et al (2009), have provided the identification of hydroxylation genes associated with reduced endosperm provitamin A content. In particular transcript profiling led to discovery of the *Hydroxylase3* locus that coincidently mapped to a carotene QTL, thereby prompting investigation of allelic variation in a broader collection. Vallabhaneni & Wurtzel (2009) have sampled a maize germplasm collection via statistical testing of the correlation between carotenoid content and candidate gene transcript levels. They observed multiple pathway bottlenecks for isoprenoid biosynthesis and carotenoid biosynthesis acting in specific temporal windows of endosperm development. Transcript levels of paralogs encoding*isoprenoid isopentenyl diphosphate* and *geranylgeranyl diphosphate*-producing enzymes, such as DXS3 (*1-deoxy-D-xylulose-5-phosphate synthase3*), *DXR* (*DXP reductoisomerase*), *HDR* (*4-hydroxy-3-methylbut-2-enyl diphosphate reductase*), and *GGPPS1* (*geranylgeranyl pyrophosphate synthase1*), were found to positively correlate with endosperm carotenoid content. Toledo-Ortiz et al. (2010) have recently identified in *Arabidopsis* seedlings that *phytochrome-interacting factor1* (*PIF1*) and other transcription factors of the *phytochrome-interacting factor* (*PIF*) family down-regulate the accumulation of carotenoids by specifically repressing the gene encoding *PSY*, the main rate-determining enzyme of the pathway. Their results also suggest a role for *PIF1* and other *PIFs* in transducing light signals to regulate *PSY* gene expression and carotenoid accumulation during daily cycles of light and dark in mature plants. In this context, manipulating the levels of *PIF* transcription factors by transgenic or marker-assisted breeding approaches might help improve carotenoid accumulation in plants for the production of varieties with enhanced agronomical, industrial, or nutritional value.

NEW STRATEGIES FOR CREATING VARIATION

The use of molecular biology to isolate, characterizes, and modifies individual genes followed by plant transformation and trait analysis will introduce new traits and more diversity into maize database. For example, maize-based diets (animals or humans) require lysine and tryptophan supplementation for adequate protein synthesis. The development of high-lysine maize to use in improved animal feeds illustrates the challenges that continually interlace metabolic engineering projects. From a biochemical standpoint, the metabolic pathway for lysine biosynthesis in plants is very similar to that in many bacteria. The key enzymes in the biosynthetic pathway are aspartakinase (AK) and dihydrodipicolinic acid synthase (DHDPS), both of which are feedback inhibited by lysine (Galili, 2004). Falco et al. (1995) isolated bacterial genes encoding lysine-insensitive forms of AK and DHDPS from *Escherichia coli* and *Corynebacterium*, respectively. A deregulated form of the plant DHDPS was created by site-specific mutagenesis (Shaver et al., 1996). The expression of the bacterial DHPS in maize seeds overproduced lysine, but they also contained higher level of lysine catabolic products then their wild-type parents (Mazur et al., 1999), despite the fact that lysine catabolism was suggested to be minimal in this tissue (Arruda et al., 2000). Likewise, a gene corresponding to a feedback-resistant form of the enzyme anthranilate synthase (AS) has been cloned from maize and re-introduced via transformation under the control of seed-specific promoters. This altered AS has reduced sensitivity to feedback inhibition by tryptophan; thus, tryptophan is overproduced and accumulates to higher than normal levels in the grain. This strategy has been successful in reaching commercially valuable levels of tryptophan in the grain (Anderson et al., 1997). More recently, Houmard et al. (2007) reported the increase in maize grains by specific suppression of lysine catabolism via RNAi. An important observation from these studies was that the lysine content was increased in the transgenic lines by 15-20% to 54.8%. These experiments showed that transgenic approaches, in addition to investigating relationships between zein synthesis and opaque endosperm, could be useful to increase kernel lysine content. Similarly, Reyes et al. (2008), using RNAi, have produced transgenic maize lines that had LKR/SDH suppressed in the embryo, endosperm or both. These authors noted a synergist increase in free lysine content in the mature kernel when LKR/SDH was suppressed in both embryo and endosperm; these results have also suggested new insights into how free lysine level is regulated and distributed in developing grains.

A different approach to enhance the level of a given amino acid in kernels is to improve the protein sink for this amino acid (Kriz, 2009). This goal can be achieved by transforming plants with genes encoding stable proteins that are rich in the desired amino acid(s) and that can accumulate to high levels. Among a variety of natural, modified or synthetic genes that were tested, the most significant increases in seed lysine levels were obtained by expressing a genetically-engineered hordothionine (HT12) or a barley high-lysine protein 8 (BHL8), containing 28 and 24% lysine, respectively (Jung and Falco, 2000). These proteins accumulated in transgenic maize to 3-6% of total grain proteins and when introduced together with a bacterial DHPS, resulted in a very high elevation of a total lysine to over 0.7% of seed dry weight

(Jung and Falco, 2000) compared to around 0.2% in wild-type maize. Similarly, Rascon-Cruz et al. (2004) have found that the introduction of a gene encoding amarantin-protein from *Amaranth* plants, which is known to be balanced in its amino acid content, increases from 8 to 44% essential amino acid content. Bicar et al. (2008) have developed transgenic maize lines that produce milk α-lactalbumin in the endosperm. They noted that the lysine content of the lines examined was 29-47% greater in endosperm from transgene positive kernels. Furthermore, Wu et al. (2007) provided a novel approach to enrich the lysine content (up to 26%) in the maize grain by endosperm-specific expression of an *Arabidopsis* lysyl tRNA synthate. Combining these traits with seed-specific reduction of lysine catabolism offers an optimistic future for commercial application of high-lysine maize.

Single mutations in starch biosynthesis have been commercially used for the production of some specialty maize. For example, specialty varieties such as waxy can result in 99% amylopectins, while the use of «amylomaize varieties» (*amylose extender* endosperm mutants) have kernels up to 20% amylopectin and 80% amylose. These varieties are of interest for commercial purposes in starch industry, such as food ingredients, sweeteners, adhesives, and for the development of thermoplastics and polyurhetanes. However, advances in understanding the starch biosynthetic pathway provide new ways to redesign starch for specific purposes, such for ethanol production. Alteration in starch structure can be achieved by modifying genes encoding the enzymes responsible for starch synthesis, many of which have more than one isoform (Boyer and Hannah, 2001). Transgenic lines with modified expression of specific starch synthases, starch branching enzymes or starch debranching enzymes are being generated in attempts to produce starch granules with increased or decreased crystallinity, and thus altered susceptibility to enzymatic digestion. Another strategy is to reduce the energy requirements for the starch to ethanol conversion process. For example, gelatinization is the first step in bioethanol production from starch. It is conceivable that a modified starch with decreased gelatinization temperature might require less energy for the conversion process. Research showed that expression of a recombinant amylopullulanase in rice resulted in starch that when heated to 85 C was completely converted into soluble sugars (Chiang, 2005). The expression of microbial genes in transgenic plants represents also an opportunity to produce renewable resources of fructans.

Transgenic maize expressing the *Bacillus amyloliquefaciens SacB* gene accumulates high-molecular weight fructose in mature seed (Caimi et al., 1996). This could potentially be exploited within the high-fructose maize syrup market. Moreover, Zhang et al. (2007) have developed transgenic maize endosperm, via the introduction of a *Streptococcus mutans gtfD* gene, that accumulates novel glucan (oligo- and polysaccharides composed solely of glucose molecules) polymers at levels relevant to commercial production. The expression of that gene yielded fully functional GTF-D enzyme as shown by accumulation of novel soluble $\alpha(1\rightarrow6)$-linked glucan at high levels in the mature maize kernels (up to 14% of their dry weight). These findings suggest that the introduction of greater diversity in linkages within α-glucan polymers will enable the generation of specialty glucans to replace modified starches used for several products (e.g. thickening reagents, adhesives, textile modification, and papermaking polymers with economic and environmental benefits).

Efforts to increase oil content and composition in maize kernels through breeding have considerable success, but high oil lines have significant reduced yield (cf Moose et al., 2004). Several and complementary approaches might be considered to try and enhance oil content in maize kernels. This goal may be achieved by increasing the relative proportion of the oil-rich embryonic tissue within the grain. It has been recently reported that embryo size and oil content could be increased in transgenic maize by ectopic expression of the wheat *Purindoline a* and *b* (*PINA* and *PINB*) genes (Zhang, et al., 2010). While total oil content of the kernel was increased by 25% in these transgenic lines, the molecular mechanism responsible for the increase remains to be clarified. If no modification of kernel size was observed in these transgenic lines, other agronomic characteristics remain to be studied to evaluate the economic potential of such material. Another strategy to increase oil accumulation in the grain may consist in improving both oil content of embryonic tissues. A close examination of C metabolism in maize embryos suggested that the flux of C through NADP-ME may constitute a metabolic bottleneck (Alonso et al., 2010). Accordingly, the oil content of the kernel was positively correlated with malic enzyme activities in maturing embryos (Doehlert & Lambert, 1991), which makes NADP-ME an attractive target for engineering high oil concentrations in embryos of maize. Furthermore, in oilseed species, numerous biotechnological approaches have been carried out that were aimed at maximizing the flow of C into oil by overexpression of enzymes of the TAG assembling network. For example in maize, several attempts have been made to over-express *diacylglycerol acyltransferases* (*DGAT*). *DGAT* catalyses the transfer of an acyl chain from the acyl-CoA pool to the*sn-3* position of a diacylglycerol molecule, resulting in the synthesis of TAG. The embryospecific over-expression of both maize *DGAT1-2* and of fungal *DGAT2* (Zheng et al., 2008; Oakes et al., 2011) resulted in limited (1.25 fold) but statistically significant increases in kernel oil content. Whereas it has been shown that grain yield was not affected by expression of fungal *DGAT2*, data concerning the putative incidence of the over-expression of maize *DGAT1-2* on yield and other agronomic characteristics of the modified lines are missing. Nevertheless these works provide insights into the molecular basis of natural variation of oil and oleic-acid contents in plants and highlight *DAGT* as a promising target for increasing oil and oleic-acid content in other crops.

The recent identification of transcriptional regulators of the oil biosynthetic network in maize has opened the way for designing and testing new original biotechnological strategies. A study has shown that seed-specific expression *ZmWRI1*, a WRI1-like gene of maize, enhanced oil accumulation in transgenic maize without detectable abnormalities. However, expression of *ZmLEC1* under similar conditions severely affected growth and development of the resulting transgenic maize plants (Shen, et al., 2010). Similar results were obtained by constitutive overexpression of the *ZmWRI1* gene in the transgenic maize plants (Pouvreau et al., 2011). It was also found that *ZmWri1* not only increases the fatty acid content of the mature maize grain but also the content of certain amino acids (Lys, Glu, Phe, Ala, Val) of several compounds involved in amino acid biosynthesis (pyro-Glu, aminoadipic acid, Orn, nor- Leu), and of two intermediates of the TCA cycle(citric acid, succinic acid) (Pouvreau et al., 2011). Finally, a third approach to increase oil content in maize grains may consist in diverting C flux from starch to oil in the endosperm. Considering both the elevated amounts of ATP consumed in futile cycling processes and the rates of reductant production in endosperm tissues of maize kernels, Alonso et al (2010) have speculated that increasing biomass synthesis and redirecting part of the C flux toward fatty acid production by metabolic engineering could theoretically be obtained. This would require inhibiting futile cycling whilst overexpressing the whole set of enzymes involved in TAG production. To date, no successful attempt has been reported. If the use of *ZmWRI1* as a biotechnological tool for improving oil content in embryos of maize seems promising (see above), over-expression of *ZmWRIa*in the starchy endosperm was not sufficient to trigger oil accumulation in this compartment (Shen et al., 2010). Since there is no evidence that WRI1 regulates TAG assembly, it is not surprising that over-expression of ZmWRI1a only proves to be efficient in tissues already accumulating oil, and thus already expressing the TAG biosynthetic machinery. What is more, the structure and size of maize kernels may impair large accumulation of oil in the endosperm.

The cloning of carotenogenic genes in maize and in other organisms have opening up the possibility of modifying and engineering the carotenoid biosynthetic pathways in plants, although question remains about the rate-controlling steps that limit the predictability of metabolic engineering in plants. Engineering high levels of specific carotenoid structures requires controlled enhancement of total carotenoid levels (enhancing pathway flux, minimizing degradation, and optimizing sequestration) plus controlled composition for specific pathway end products. While most of the nuclear genes for the plastid-localized pathway are available (Li et al. 2007) and/or can be identified, questions remain about the rate-controlling steps that limit the predictability of metabolic engineering in plants. Predictable manipulation of the seed carotenoid biosynthetic pathway in diverse maize genotypes necessitates the elucidation of biosynthetic step(s) that control carotenoid accumulation in endosperm tissue. Studies have implicated PSY, the first committed enzyme, as rate controlling for endosperm carotenoids (e.g.Pozniak et al., 2007; Li et al., 2008). However, upstream precursor pathways may also positively influence carotenoid accumulation (Matthews and Wurtzel, 2000; Mahmoud & Croteau, 2001), while downstream degradative pathways may deplete the carotenoid pool (Galpaz et al., 2008).

Transgenic strategies can also be used as tools to complement breeding techniques in meeting the estimated levels of provitamin A. In this respect, Aluru et al. (2008) reported that the overexpression of the bacterial genes *crtB* (for PS) and *crtI* (for the four desaturation steps of the carotenoid pathway catalyzed by PDS and β-carotene desaturase in plants), resulted in an increase of total carotenoids of up to 34-fold with a preferential accumulation of β-carotene in the maize endosperm. The levels attained approach those estimated to have a significant impact on the nutritional status of target populations in developing countries. Furthermore, the same authors, via gene expression analyses, suggested that increased accumulation of β -carotene is due to an up-regulation of the endogenous lycopene β -cylase. These experiments set the stage for the design of transgenic approaches to generate provitamin A-rich maize that will help alleviate vitamin A deficiency in developing countries. Similarly, Naqvi et al. (2009) produced transgenic maize plants with significantly increased contents for β -carotene, ascorbate, and folate in the endosperm via that simultaneous modification of 3 separate metabolic pathways. The transgenic kernels contained 169-fold the normal amount of β -carotene, 6-fold, and 2-fold the normal amount of ascorbate and folate, respectively. This finding, which largely exceeds any realized thus far by conventional breeding alone, opens the way for the development of nutritional complete cereals to benefits the consumers in developing countries. Moreover, this is a very important proof of concept for genetic manipulation of distinct metabolic pathways.

There is evidence indicating that tocophenols, in particular γ -tocophenol the predominant form of vitamine E in plant seeds, are indispensable for protection of the polyunsaturated fatty acid in addition to have benefits to the meet industry (Rocheford et al., 2002). The same authors have also shown that considerable variation is present among different maize inbreds from tocophenol levels, as well as different ratios of α-tocophenol to γ-tocophenol. This result suggested that breeders can use natural varieties, molecular marker assisted selection strategies and transgenic technologies to alter overall level of tocophenols and ratio of α- to γ-tocophenol. However, current nutritional research on the relative and unique benefits of α- to γ-tocophenol should be considered in developing breeding strategies to alter levels of these vitamin E compounds.

Another area in which transgenic approaches may help solve an important problem with maize as a feed grain is in the reduction of phytic acid levels. In maize, 80% of the total phosphorous (P) is found as phytic acid, and most of that is in the germ (O'Dell et al., 1972). Phytate P is very poorly digested by non-ruminant animals, therefore inorganic supplementation is necessary. Phytate is also a strong chelator that reduces the bioavailability of several other essential minerals such as Ca, Zn, Cu, Mn, and Fe. In addition, since the phytate in the diet is poorly digested, the excrement of monogastric animals (e.g. poultry and pigs), is rich in P and this contributes significantly to environmental pollution. *Low phytic acid* mutants (*lpa*) of maize are available; these have received considerable attention by breeders in order to develop commercially acceptable hybrids with reduced levels of phytic acid (Raboy, 2009).

In maize, several mutants with low levels of phytate have been isolated and mapped; this includes *lpa 1-1*, *lpa 2-1,* and *lpa 241*, (Raboy, 2009). The *lpa1* mutant does not accumulate *myo*-inositol monophosphate or polyphosphate intermediates. It has been proposed that *lpa1* is a mutation in *myo*-inositol supply, the first part of the phytic acid biosynthesis pathway (Raboy et al., 2000). The *lpa2*mutant has reduced phytic acid content in seeds and accumulates *myo*-inositol phosphate intermediates. Maize *lpa2* gene encodes a *myo*-inositol phosphate kinase that belongs to the Ins(1,3,4)P3 5/6-kinase gene family (Shi et al., 2003). The *lpa3* mutant seeds have reduced phytic acid content and accumulate*myo*-inositol, but not *myo*-inositol phosphate intermediates was found to encode *myo*-inositol kinase (Shi et al., 2005).

Despite efforts to elucidate and manipulate phytic acid biosynthesis, *low phytic acid* mutants have limited value to breeders because of adverse effects on agronomic traits such as low germination rates, reduced seed weight (*lpa1-1*), stunted vegetative growth and impaired seed development (*lpa241*). However, Shi et al. (2007) have recently identified the gene disrupted in maize *lpa1* mutants as a multidrug resistance-associated protein (MRP) ATP-binding cassette (ABC) transporter. Silencing expression of this transporter using the embryo-specific globuline promoter produced low-phytic acid, high phosphate transgenic maize seeds that germinate normally and do not show any significant reduction in seed dry weight.

To increase the amount of bioavailable iron in maize, Drakakaki et al. (2005) have generated transgenic maize plants expressing aspergillus phytase and iron-binding protein ferritin. This strategy has proven effective for increasing iron availability and enhancing its absorption. However, much work is still to be done to transfer this technology to tropical and subtropical maize genotypes normally grown in the areas of greatest need for enhanced iron content maize.

A relatively new area in plant biotechnology is the use of genetically-engineered maize to produce high-value end products such as vaccines, therapeutic proteins, industrial enzymes and specialty chemicals (see Hood & Howard, 2009 for a review). The long-term commercial expectations for this use of "plants as factories", often also called "molecular farming", are large. Transgenic maize seed has many attractive features for this purpose, including: i) well-suited for the production and storage of recombinant proteins; ii) ease of scale-up to essentially an infinite capacity; iii) well-established infrastructure for producing, harvesting, transporting, storing, and processing; iv) low cost of production; v) freedom from animal pathogenic contaminants; vi) relative ease of producing transgenic plants which express foreign proteins of interest. However, there is a need, apart the public issues related with the acceptance of genetically-engineered maize, for continued efforts in increasing expression in order to reduce cost effectiveness for products at protein accumulation levels in transgenic plants to broaden this new uses.

Conclusion and Future Perspectives

Two prominent features of agriculture in the 20th century have been the use of breeding and genetics to boost crop productivity and the use of agricultural chemicals to protect crops and enhance plant growth. In the 21st century, crops must produce good yields while conserving land, water, and labor resources. At the same time, industries and consumers require plants with an improved and novel variation in grain composition. We expect that genomics will bolster plant biochemistry as researchers seek to understand the metabolic pathways for the synthesis of these compounds. Identifying rate-limiting steps in synthesis could provide targets for genetically engineering biochemical pathways to produce augmented amounts of compounds and new compounds. Targeted expression will be used to channel metabolic flow into new pathways, while gene-silencing tools will reduce or eliminate undesirable compounds or traits. Therefore, developing plants with improved grain quality traits involves overcoming a variety of technical challenges inherent in metabolic engineering programs.

Metabolism is one of the most important and best recognized networks within biological systems. However, advances in the understanding of metabolic regulation still suffer from insufficient research concerning the modular operation of such networks. Elucidation of metabolic regulation within the context of the entire system, including transcriptional, translational and posttranslational mechanisms, is rarely attempted (Sweetlove et al., 2008). Instead, to date, studies on metabolic regulation have mostly been limited to regulatory interactions within the metabolic pathways themselves (Sweetlove & Fernie, 2005; Sweetlove et al., 2008). Strategies to detect intermediary metabolic fluxes can now be estimated by computer-aided modeling of the central metabolic network and by mapping the pattern of metabolic fluxes underlying, via the possibility of labeling data collected by NMR and GC-MS and the biomass composition. For example Alonso et al., (2011), to map the pattern of metabolic fluxes underlying this efficiency, have labeled maize embryos to isotopic steady state using a combination of labeled ^{13}C-substrates. The resultant flux map reveals that even though 36% of the entering C goes through the oxidative pentose-phosphate pathway; this does not fully meet the NADPH demands for fatty acid synthesis. Metabolic flux analysis and enzyme activities have highlighted the importance of plastidic NADP-dependent malic enzyme, which provides one-third of the C and NADPH required for fatty acid synthesis in developing maize embryos.

It should worth to be mentioned that metabolic engineering of maize has been relatively slow due to the difficulty of maize transformation. Maize transformation with *Agrobacterium* is now more efficient than currently used particle gun transformation (reviewed in Jones, 2009; Reyes et al., 2010). In addition, larger DNA fragments can be inserted with *Agrobacterium* than those previously reported by other methods. The ability to routinely insert metabolic pathway quantities of DNA into the maize genome will further speed up maize metabolic engineering. Furthermore, site-directed mutagenesis via gene targeting, based on homologous recombination such as the application of designed zinc finger nucleases that induce a double stranded break at their target locus, are promising tools for genetic applications (Shukla et al., 2009; Saika et al., 2011). The use of these technologies may lead to both targeted mutagenesis and targeted gene replacement at remarkably high frequencies and enable to modify useful information, acquired from structural- and computational-based protein engineering, to be applied directly to molecular breeding of crops, including metabolic engineering.

Advances in plant genetics and genomic technologies are also contributing to the acceleration of gene discovery for maize product development. In the past few years there has been much progress in the development of strategies to discover new plant genes. In large part, these developments derive from four experimental approaches: firstly, genetic and physical mapping in plants and the associated ability to use map-based gene isolation strategies; secondly, transposon tagging which allows the direct isolation of a gene via forward and reverse genetic strategies as well as the development of the Targeting Induced Local Lesions IN Genomes (TILLING) technique; thirdly, protein-protein interaction cloning, that permits the isolation of multiple genes contributing to a single pathway or metabolic process. Finally, through bioinformatics/genomics, the development and use of large ESTs databases (http://www.maizegdb.org) and, DNA microarray technology to investigate mRNA-level controls of complex pathways. Moreover, new technologies and information continue to increase our understanding of maize; for instance, the complete DNA sequence of the maize genome, along with comprehensive transcriptome, proteome, metabolome, and epigenome information, is also a key resource for advancing fundamental knowledge of the biology of development seed quality-related traits to be applied in molecular breeding and biotechnology. These additional layers of information should help to further unravel the complexities of how genes and gene networks function to give plants including quality-traits. This knowledge will drive to improved predictions and capacities to assemble gene variation through molecular breeding as well as more optimal gene selection and regulation in the development of future biotechnology products.

In conclusion, although, conventional breeding, molecular marker assisted breeding, and genetic engineering have already had, and will continue to have, important roles in maize improvement, the rapidly expanding information from genomics and genetics combined with improved genetic engineering technology offer a wide range of possibilities for the improvement of the maize grain.

ACKNOWLEDGEMENTS

We apologize to all those whose contributions we did not include in this review because of space constraints, personal preferences, or simple oversight. In addition, we would like to thank the members of the laboratory for their research contributions which are described here. The work was supported by Ministero per le Politiche Agricole, Alimentari e Forestali, Rome.

REFERENCES

1. P. A. Alonso, D. L. Val, Y. Shachar-Hill, 2010 Understanding fatty acid synthesis in developing maize embryos using metabolic flux analysis Metabolic Engineering, 12 5 488 497
2. A. P. Alonso, D. L. Val, Y. Shachar-Hill, 2011 Central metabolic fluxes in the endosperm of developing maize seeds and their implications for metabolic engineering Metabolic Engineering, 13 1 96 107
3. M. Aluru, Y. Xu, R. Guo, Z. Wang, S. Li, W. White, K. Wang, S. Rodermel, 2008 Generation of transgenic maize with enhanced provitamin A content Journal of Experimental Botany, 59 13 3551 3562
4. P. C. Anderson, P. S. Chomet, M. C. Griffor, A. L. Kriz, 1997 Anthranilate synthase gene and method of use thereof for conferring tryptophan overproduction
5. P. Arruda, E. L. Kemper, F. Papes, A. Leite, 2000 Regulation of lysine catabolism in higher plants Trends in Plant Science, 5 8 324 330
6. R. A. Azevedo, P. Arruda, W. L. Turner, P. J. Lea, 1997 The biosynthesis and metabolism of the aspartate derived amino acids in higher plants Phytochemistry, 46 3 395 419
7. E. H. Bicar, W. Woodman-Clikeman, V. Sangtong, J. M. Peterson, S. S. Yang, M. Lee, M. P. Scott, 2007 Transgenic maize endosperm containing a milk protein has improved amino acid balance Transgenic Research, 17 1 59 71
8. P. Botella-Pavía, M. Rodríguez-Concepción, 2006 Carotenoid

biotechnology in plants for nutritionally improved foods Physiologia Plantarum, 126 3 369 381

9. C. Boyer, L. Hannah, 2001 Kernel mutants of Corn, In: Specialty Corns. 2nd ed., A.R. Hallauer, 1 32 , C.R.C. Press, Boca Raton.

10. O. V. Brenna, N. Berardo, 2004 Application of Near-Infrared Reflectance Spectroscopy (NIRS) to the Evaluation of Carotenoids Content in Maize Journal of Agricultural and Food Chemistry, 52 18 5577 5582

11. M. R. Brochetto-Braga, A. Leite, P. Arruda, 1992 Partial Purification and Characterization of Lysine-Ketoglutarate Reductase in Normal and Opaque-2 Maize Endosperms Plant Physiology, 98 3 1139 1147

12. P. G. Caimi, L. M. Mc Cole, T. M. Klein, P. S. Kerr, 1996 Fructan Accumulation and Sucrose Metabolism in Transgenic Maize Endosperm Expressing a Bacillus amyloliquefaciens SacB Gene Plant Physiology, 110 2 355 363

13. S. Chander, Y. Q. Guo, X. H. Yang, J. Zhang, X. Q. Lu, J. B. Yan, T. M. Song, T. R. Rocheford, J. S. Li, 2007 Using molecular markers to identify two major loci controlling carotenoid contents in maize grain Theor. Appl. Genet., 116 2 223 233

14. Chih. Chiang-Ming, Feng. Yeh-Shi, Li. Huang-Fen, Tung. Tseng-Hi, Mei. Chung-Chu, Chang. Wang-Sheng, Hu. Lur-Shen, Jei. Shaw-Fu, Su. Yu-May, 2005 Expression of a bi-functional and thermostable amylopullulanase in transgenic rice seeds leads to autohydrolysis and altered composition of starch Molecular Breeding, 15 2 125 143

15. D. Doehlert, R. Lambert, 1991 Metabolic characteristics associated with starch, protein, and oil deposition in developing maize kernels Crop Sci., 31 151 157

16. G. Drakakaki, S. Marcel, R. P. Glahn, E. K. Lund, S. Pariagh, R. Fischer, P. Christou, E. Stoger, 2005 Endosperm-Specific Co-Expression of Recombinant Soybean Ferritin and Aspergillus Phytase in Maize Results in Significant Increases in the Levels of Bioavailable Iron Plant Molecular Biology, 59 6 869 880

17. J. Dudley, R. Lambert, 1992 Ninety generations of selection for oil and protein in maize Maydica, 37 1 7

18. S. Falco, C., T. Guida, M. Locke, J. Mauvais, C. Sanders, R. T.

Ward, P. Webber, 1995 Transgenic Canola and Soybean Seeds with Increased Lysine Nat Biotech, 13 6 577 582

19. P. D. Fraser, P. M. Bramley, 2004 The biosynthesis and nutritional uses of carotenoids Progress in Lipid Research, 43 3 228 265
20. F. Fu-F, H. Xue-W, 2010 Coexpression Analysis Identifies Rice Starch Regulator1, a Rice AP2/EREBP Family Transcription Factor, as a Novel Rice Starch Biosynthesis Regulator Plant Physiology, 154 2 927 938
21. G. Galili, 2004 New insights into the regulation and functional significance of lysine metabolism in plants Ann. Rev. Plant Biol., 53 1 27 43
22. N. Galpaz, Q. Wang, N. Menda, D. Zamir, J. Hirschberg, 2008 Abscisic acid deficiency in the tomato mutant high-pigment 3 leading to increased plastid number and higher fruit lycopene content The Plant Journal, 53 5 717 730
23. B. C. Gibbon, B. A. Larkins, 2005 Molecular genetic approaches to developing quality protein maize Trends in Genetics, 21 4 227 233
24. M. J. Giroux, J. Shaw, G. Barry, B. G. Cobb, T. Greene, T. Okita, L. C. Hannah, 1996 A single mutation that increases maize seed weight Proceedings of the National Academy of Sciences, 93 12 5824 5829
25. S. Gygi, Y. Rochon, B. Franza, R. Aebersold, 1999 Correlation between protein and mRNA abundance in yeast Molecular and cellular biology, 19 3 1720
26. L. C. Hannah, 2007 Starch Formation in the Cereal Endosperm, In: Endosperm, Olsen, Odd Arne, 179 193 , Springer Berlin Heidelberg
27. L. Hannah, M. James, 2008 The complexities of starch biosynthesis in cereal endosperms Current Opinion in Biotechnology, 19 2 160 165
28. C. E. Harjes, T. R. Rocheford, L. Bai, T. P. Brutnell, C. B. Kandianis, S. G. Sowinski, A. E. Stapleton, R. Vallabhaneni, M. Williams, E. T. Wurtzel, J. Yan, E. S. Buckler, 2008 Natural Genetic Variation in Lycopene Epsilon Cyclase Tapped for Maize Biofortification Science, 319 5861 330 333

29. H. Hartings, M. Lauria, N. Lazzaroni, R. Pirona, M. Motto, 2011 The Zea mays mutants opaque-2 and opaque-7 disclose extensive changes in endosperm metabolism as revealed by protein, amino acid, and transcriptome-wide analyses BMC genomics, 12 1 41

30. D. R. Holding, B. G. Hunter, T. Chung, B. C. Gibbon, C. F. Ford, A. K. Bharti, J. Messing, B. R. Hamaker, B. A. Larkins, 2008 Genetic analysis of opaque2 modifier loci in quality protein maize Theoretical and Applied Genetics, 117 2 157 170

31. D. R. Holding, B. A. Larkins, 2009 Zein Storage Proteins, In: Molecular Genetic Approaches to Maize Improvement, Kriz, Alan L. and Larkins, Brian A., 269 286 , Springer Berlin Heidelberg

32. D. R. Holding, R. B. Meeley, J. Hazebroek, D. Selinger, F. Gruis, R. Jung, B. A. Larkins, 2010 Identification and characterization of the maize arogenate dehydrogenase gene family Journal of Experimental Botany, 61 13 3663 3673

33. E. Hood, J. Howard, 2009 Over-expression of novel proteins in maize, In: Molecular genetic approaches to maize improvement, A.L. Kriz, B.A. Larkins, 91 105 , SpringerVerlag, Berlin Heidelberg.

34. N. M. Houmard, J. L. Mainville, C. P. Bonin, S. Huang, M. H. Luethy, T. M. Malvar, 2007 High-lysine corn generated by endosperm-specific suppression of lysine catabolism using RNAi Plant Biotechnology Journal, 5 5 605 614

35. S. Huang, W. R. Adams, Q. Zhou, K. P. Malloy, D. A. Voyles, J. Anthony, A. L. Kriz, M. H. Luethy, 2004 Improving Nutritional Quality of Maize Proteins by Expressing Sense and Antisense Zein Genes Journal of Agricultural and Food Chemistry, 52 7 1958 1964

36. B. G. Hunter, M. K. Beatty, G. W. Singletary, B. R. Hamaker, B. P. Dilkes, B. A. Larkins, R. Jung, 2002 Maize Opaque Endosperm Mutations Create Extensive Changes in Patterns of Gene Expression The Plant Cell Online, 14 10 2591 2612

37. P. C. L. John, 2007 Hormonal Regulation of Cell Cycle Progression and its Role in Development, In: Cell cycle control and plant development, Inze D., 311 334 , Blackwell, Oxford

38. R. Jung, S. Falco, 2000 Transgenic corn with an improved amino acid composition 8th Intl. Sump. on Plant Seeds, Gatersleben, Germany

39. K. Koch, 2004 Sucrose metabolism: regulatory mechanisms and pivotal roles in sugar sensing and plant development Current Opinion in Plant Biology, 7 3 235 246
40. Alan. L. Kriz, 2009 Enhancement of Amino Acid Availability in Corn Grain, In: Molecular Genetic Approaches to Maize Improvement, Kriz, Alan L. and Larkins, Brian A., 79 89 , Springer Berlin Heidelberg
41. A. Kubo, C. Colleoni, J. R. Dinges, Q. Lin, R. R. Lappe, J. G. Rivenbark, A. J. Meyer, S. G. Ball, M. G. James, T. A. Hennen-Bierwagen, A. M. Myers, 2010 Functions of Heteromeric and Homomeric Isoamylase-Type Starch-Debranching Enzymes in Developing Maize Endosperm Plant Physiology, 153 3 956 969
42. A. C. Kurilich, J. A. Juvik, 1999 Quantification of Carotenoid and Tocopherol Antioxidants in Zea mays Journal of Agricultural and Food Chemistry, 47 5 1948 1955
43. J. Lai, N. Dey, Cheol. Kim-Soo, A. K. Bharti, S. Rudd, K. F. X. Mayer, B. A. Larkins, P. Becraft, J. Messing, 2004 Characterization of the Maize Endosperm Transcriptome and Its Comparison to the Rice Genome Genome Research, 14 10a 1932 1937
44. R. Lambert, (2001) High-oil corn hybrids, In: Specialty Corns. 2nd ed., Hallauer, A.R., pp. pp 131-154, C.R.C. Press, Boca Raton.
45. C. C. Laurie, S. D. Chasalow, J. R. Le Deaux, R. Mc Carroll, D. Bush, B. Hauge, C. Lai, D. Clark, T. R. Rocheford, J. W. Dudley, 2004 The Genetic Architecture of Response to Long-Term Artificial Selection for Oil Concentration in the Maize Kernel Genetics, 168 4 2141 2155
46. J. W. Lawton, 2002 Zein: A History of Processing and Use Cereal Chemistry, 79 1 1 18
47. C. Lending, B. Larkins, 1989 Changes in the zein composition of protein bodies during maize endosperm development The Plant Cell Online, 1 10 1011
48. F. Li, C. Murillo, E. T. Wurtzel, 2007 Maize Y9 Encodes a Product Essential for 15-cis-ζ-Carotene Isomerization Plant Physiology, 144 2 1181 1189
49. F. Li, R. Vallabhaneni, J. Yu, T. Rocheford, E. Wurtzel, 2008 The maize phytoene synthase gene family: overlapping roles

for carotenogenesis in endosperm, photomorphogenesis, and thermal stress tolerance Plant physiology, 147 3 1334

50. X. Liu, J. Fu, D. Gu, W. Liu, T. Liu, Y. Peng, J. Wang, G. Wang, 2008 Genome-wide analysis of gene expression profiles during the kernel development of maize (Zea mays L.) Genomics, 91 4 378 387
51. I. Llop-Tous, S. Madurga, E. Giralt, P. Marzabal, M. Torrent, M. D. Ludevid, 2010 Relevant Elements of a Maize γ-Zein Domain Involved in Protein Body Biogenesis Journal of Biological Chemistry, 285 46 35633 35644
52. S. Locatelli, P. Piatti, M. Motto, V. Rossi, 2009 Chromatin and DNA Modifications in the Opaque2-Mediated Regulation of Gene Transcription during Maize Endosperm Development The Plant Cell Online, 21 5 1410 1427
53. K. Maeo, T. Tokuda, A. Ayame, N. Mitsui, T. Kawai, H. Tsukagoshi, S. Ishiguro, K. Nakamura, 2009 An AP2-type transcription factor, WRINKLED1, of Arabidopsis thaliana binds to the AW-box sequence conserved among proximal upstream regions of genes involved in fatty acid synthesis The Plant Journal, 60 3 476 487
54. S. Mahmoud, R. Croteau, 2001 Metabolic engineering of essential oil yield and composition in mint by altering expression of deoxyxylulose phosphate reductoisomerase and menthofuran synthase Proceedings of the National Academy of Sciences, 98 15 8915
55. P. Matthews, E. Wurtzel, 2000 Metabolic engineering of carotenoid accumulation in Escherichia coli by modulation of the isoprenoid precursor pool with expression of deoxyxylulose phosphate synthase Applied microbiology and biotechnology, 53 4 396 400
56. P. Matthews, E. Wurtzel, 2007 Biotechnology of food colorant production., In: Food colorants: chemical and functional properties, C. Socaciu, 347 398 , CRC Press: Boca Raton, FL
57. B. Mazur, E. Krebbers, S. Tingey, 1999 Gene discovery and product development for grain quality traits Science, 285 5426 372 375
58. M. Miclaus, J. Xu, J. Messing, 2011 Differential gene expression and epiregulation of alpha zein gene copies in maize haplotypes. PLoS Genetics, 7 6 e1002131

59. M. Miclaus, Y. Wu, Y. Xu, H. Dooner, J. Messing, 2011 Opaque7 encodes an acyl-CoA synthetase-like protein in maize. Genetics (in press)

60. S. P. Moose, J. W. Dudley, T. R. Rocheford, 2004 Maize selection passes the century mark: a unique resource for 21st century genomics Trends in Plant Science, 9 7 358 364

61. M. Motto, C. Balconi, H. Hartings, M. Lauria, V. Rossi, 2009 Improvment of quality-related traits in maize grain: gene identification and exploitation Maydica, 54 2 3, 321 342

62. A. Myers, M. James, Q. Lin, G. Yi, P. Stinard, T. Hennen-Bierwagen, P. Becraft, 2011 Maize opaque5 encodes monogalactosyldiacylglycerol synthase and specifically affects galactolipids necessary for amyloplast and chloroplast function Plant Cell,,23 6 2331 2347

63. A. M. Myers, M. K. Morell, M. G. James, S. G. Ball, 2000 Recent Progress toward Understanding Biosynthesis of the Amylopectin Crystal Plant Physiology, 122 4 989 998

64. V. Méchin, T. Balliau, S. Château-Joubert, M. Davanture, O. Langella, L. Négroni, Jean. Prioul-Louis, C. Thévenot, M. Zivy, C. Damerval, 2004 A two-dimensional proteome map of maize endosperm Phytochemistry, 65 11 1609 1618

65. S. Naqvi, C. Zhu, G. Farre, K. Ramessar, L. Bassie, J. Breitenbach, Conesa. D. Perez, G. Ros, G. Sandmann, T. Capell, P. Christou, 2009 Transgenic multivitamin corn through biofortification of endosperm with three vitamins representing three distinct metabolic pathways Proceedings of the National Academy of Sciences, 106 19 7762 7767

66. B. L. O'Dell, A. R. De Boland, S. R. Koirtyohann, 1972 Distribution of phytate and nutritionally important elements among the morphological components of cereal grains Journal of Agricultural and Food Chemistry, 20 3 718 723

67. J. Oakes, D. Brackenridge, R. Colletti, M. Daley, D. J. Hawkins, H. Xiong, J. Mai, S. E. Screen, D. Val, K. Lardizabal, K. Gruys, J. Deikman, 2011 Expression of Fungal diacylglycerol acyltransferase2 Genes to Increase Kernel Oil in Maize Plant Physiology, 155 3 1146 1157

68. B. Pouvreau, S. Baud, V. Vernoud, V. Morin, C. Py, G. Gendrot, Jean. Pichon-Philippe, J. Rouster, W. Paul, P. M. Rogowsky, 2011 Duplicate Maize Wrinkled1 Transcription Factors Activate Target Genes Involved in Seed Oil Biosynthesis Plant Physiology, 156 2 674 686

69. C. Pozniak, R. Knox, F. Clarke, J. Clarke, 2007 Identification of QTL and association of a phytoene synthase gene with endosperm colour in durum wheat TAG Theoretical and Applied Genetics, 114 3 525 537

70. J. L. Prioul, V. Méchin, P. Lessard, C. Thévenot, M. Grimmer, S. Chateau-Joubert, S. Coates, H. Hartings, M. Kloiber-Maitz, A. Murigneux, X. Sarda, C. Damerval, K. J. Edwards, 2008 A joint transcriptomic, proteomic and metabolic analysis of maize endosperm development and starch filling Plant Biotechnology Journal, 6 9 855 869

71. V. Raboy, P. F. Gerbasi, K. A. Young, S. D. Stoneberg, S. G. Pickett, A. T. Bauman, P. P. Murthy, N., W. F. Sheridan, D. S. Ertl, 2000 Origin and Seed Phenotype of Maize low phytic acid 1-1 and low phytic acid 2-1 Plant Physiology, 124 1 355 368

72. V. Raboy, 2009 Seed Total Phosphate and Phytic Acid, In: Molecular Genetic Approaches to Maize Improvement, Kriz, Alan L. and Larkins, Brian A., 41 53 , Springer Berlin Heidelberg

73. Q. Rascon-Cruz, S. Sinagawa-Garcia, J. Osuna-Castro, N. Bohorova, O. Paredes-López, 2004 Accumulation, assembly, and digestibility of amarantin expressed in transgenic tropical maize TAG Theoretical and Applied Genetics, 108 2 335 342

74. A. R. Reyes, C. P. Bonin, N. M. Houmard, S. Huang, T. M. Malvar, 2008 Genetic manipulation of lysine catabolism in maize kernels Plant Molecular Biology, 69 1-2 , 81 89

75. F. Reyes, B. Sun, H. Guo, D. Gruis, M. Otegui, 2010 Agrobacterium tumefaciens-mediated transformation of maize endosperm as a tool to study endosperm cell biology Plant physiology, 153 2 624

76. T. Rocheford, J. Wong, C. Egesel, R. Lambert, 2002 Enhancement of Vitamin E Levels in Corn J. Am. College Nutr., 21 191 198

77. H. Saika, A. Oikawa, F. Matsuda, H. Onodera, K. Saito, S. Toki, 2011 Application of gene targeting to designed mutation breeding of high-tryptophan rice Plant Physiology, 156 3 1269

78. G. Sandmann, S. Römer, P. D. Fraser, 2006 Understanding carotenoid metabolism as a necessity for genetic engineering of crop plants Metabolic Engineering, 8 4 291 302

79. G. Segal, R. Song, J. Messing, 2003 A New Opaque Variant of Maize by a Single Dominant RNA-Interference-Inducing Transgene Genetics, 165 1 387 397

80. J. M. Shaver, D. C. Bittel, J. M. Sellner, D. A. Frisch, D. A. Somers, B. G. Gengenbach, 1996 Single-amino acid substitutions eliminate lysine inhibition of maize dihydrodipicolinate synthase. Proceedings of the National Academy of Sciences of the United States of America, 93 5 1962 1966

81. B. Shen, W. B. Allen, P. Zheng, C. Li, K. Glassman, J. Ranch, D. Nubel, M. C. Tarczynski, 2010 Expression of ZmLEC1 and ZmWRI1 Increases Seed Oil Production in Maize Plant Physiology, 153 3 980 987

82. J. Shi, H. Wang, K. Schellin, B. Li, M. Faller, J. M. Stoop, R. B. Meeley, D. S. Ertl, J. P. Ranch, K. Glassman, 2007 Embryo-specific silencing of a transporter reduces phytic acid content of maize and soybean seeds Nat Biotech, 25 8 930 937

83. V. K. Shukla, Y. Doyon, J. C. Miller, R. C. De Kelver, E. A. Moehle, S. E. Worden, J. C. Mitchell, N. L. Arnold, S. Gopalan, X. Meng, V. M. Choi, J. M. Rock, Y. Y. Wu, G. E. Katibah, G. Zhifang, D. Mc Caskill, Simpson, B. Blakeslee, S. A. Greenwalt, H. J. Butler, S. J. Hinkley, L. Zhang, E. J. Rebar, P. D. Gregory, F. D. Urnov, 2009 Precise genome modification in the crop species Zea mays using zinc-finger nucleases Nature, 459 7245 437 441

84. S. M. Sousa, M. d. G. Paniago, P. Arruda, J. A. Yunes, 2008 Sugar levels modulate sorbitol dehydrogenase expression in maize Plant Molecular Biology, 68 3 203 213

85. D. Sun, P. Gregory, C. O. Grogan, 1978 Inheritance of saturated fatty acids in maize Journal of Heredity, 69 5 341 342

86. L. Sweetlove, D. Fell, A. Fernie, 2008 Getting to grips with the plant metabolic network Biochem. J, 409 27 41

87. L. Sweetlove, A. Fernie, 2005 Regulation of metabolic networks: understanding metabolic complexity in the systems biology era New Phytologist, 168 1 9 24

88. L. D. Val, S. H. Schwartz, M. R. Kerns, J. Deikman, 2009 Development of a High Oil Trait for Maize, In: Molecular Genetic Approaches to Maize Improvement, Kriz, Alan L. and Larkins, Brian A., 303 323 , Springer Berlin Heidelberg

89. R. Vallabhaneni, C. E. Gallagher, N. Licciardello, A. J. Cuttriss, R. F. Quinlan, E. T. Wurtzel, 2009 Metabolite Sorting of a Germplasm Collection Reveals the Hydroxylase3 Locus as a New Target for Maize Provitamin A Biofortification Plant Physiology, 151 3 1635 1645

90. R. Vallabhaneni, E. T. Wurtzel, 2009 Timing and Biosynthetic Potential for Carotenoid Accumulation in Genetically Diverse Germplasm of Maize Plant Physiology, 150 2 562 572

91. N. C. Verza, T. R. Silva, G. C. Neto, F. T. S. Nogueira, P. H. Fisch, V. E. Rosa, M. M. Rebello, A. L. Vettore, F. R. Silva, P. Arruda, 2005 Endosperm-preferred Expression of Maize Genes as Revealed by Transcriptome-wide Analysis of Expressed Sequence Tags Plant Molecular Biology, 59 2 363 374

92. Y. Wan, R. Poole, A. Huttly, C. Toscano-Underwood, K. Feeney, S. Welham, M. Gooding, C. Mills, K. Edwards, P. Shewry, others 2008 Transcriptome analysis of grain development in hexaploid wheat BMC genomics, 9 1 121

93. G. Wang, X. Sun, G. Wang, F. Wang, Q. Gao, X. Sun, Y. Tang, C. Chang, J. Lai§, L. Zhu, Z. Xu, R. Song, 2011 Opaque7 encodes an AAE3 -like protein that affects storage protein synthesis in maize endosperm Genetics (in press)

94. Z. Wang, X. Chen, J. Wang, T. Liu, Y. Liu, L. Zhao, G. Wang, 2007 Increasing maize seed weight by enhancing the cytoplasmic ADP-glucose pyrophosphorylase activity in transgenic maize plants Plant Cell, Tissue and Organ Culture, 88 1 83 92

95. H. Washida, A. Sugino, S. Kaneko, N. Crofts, C. Sakulsingharoj, D. Kim, Sang. Choi-Bong, S. Hamada, M. Ogawa, C. Wang, A. Esen, T. J. V. Higgins, T. W. Okita, 2009 Identification of cis-localization elements of the maize 10-kDa delta-zein and their use in targeting RNAs to specific cortical endoplasmic reticulum subdomains The Plant Journal: For Cell and Molecular Biology, 60 1 146 155

96. P. J. White, L. A. Johnson, 2003 Corn: chemistry and technology., American Association of Cereal Chemists

97. X. R. Wu, A. Kenzior, D. Willmot, S. Scanlon, Z. Chen, A. Topin, S. H. He, A. Acevedo, W. R. Folk, 2007 Altered expression of plant lysyl tRNA synthetase promotes tRNA misacylation and translational recoding of lysine The Plant Journal, 50 4 627 636

98. Y. Wu, D. Holding, J. Messing, 2010 $-Zeins are essential for endosperm modification in quality protein maize Proceedings of the National Academy of Sciences, 107 29 12810

99. Y. Wu, J. Messing, 2011 Novel Genetic Selection System for Quantitative Trait Loci of Quality Protein Maize Genetics

100. Y. Wu, J. Messing, 2010 RNA Interference-Mediated Change in Protein Body Morphology and Seed Opacity through Loss of Different Zein Proteins Plant Physiology, 153 1 337 347

101. E. T. Wurtzel, T. R. Rocheford, J. C. Wong, R. J. Lambert, 2004 QTL and candidate genes phytoene synthase and ?-carotene desaturase associated with the accumulation of carotenoids in maize TAG Theoretical and Applied Genetics, 108 2 349 359

102. J. Yan, C. B. Kandianis, C. E. Harjes, L. Bai, Eun. Kim-Ha, X. Yang, D. J. Skinner, Z. Fu, S. Mitchell, Q. Li, M. G. S. Fernandez, M. Zaharieva, R. Babu, Y. Fu, N. Palacios, J. Li, D. Della Penna, T. Brutnell, E. S. Buckler, M. L. Warburton, T. Rocheford, 2010 Rare genetic variation at Zea mays crtRB1 increases [beta]-carotene in maize grain Nat Genet, 42 4 322 327

103. J. Yu, P. Peng, X. Zhang, Q. Zhao, D. Zhy, X. Sun, J. Liu, G. Ao, 2004 Seed-specific expression of a lysine rich protein sb401 gene significantly increases both lysine and total protein content in maize seeds Molecular Breeding, 14 1 1 7

104. J. Zhang, J. M. Martin, B. Beecher, C. Lu, L. C. Hannah, M. L. Wall, I. Altosaar, M. J. Giroux, 2010 The ectopic expression of the wheat Puroindoline genes increase germ size and seed oil content in transgenic corn Plant Molecular Biology, 74 4-5 , 353 365

105. S. Zhang, J. Dong, T. Wang, S. Guo, K. Glassman, J. Ranch, S. Nichols, 2007 High level accumulation of $-glucan in maize kernels by expressing the gtfD gene from Streptococcus mutans Transgenic research, 16 4 467 478

106. P. Zheng, W. B. Allen, K. Roesler, M. E. Williams, S. Zhang, J. Li, K. Glassman, J. Ranch, D. Nubel, W. Solawetz, D. Bhattramakki,

V. Llaca, S. Deschamps, Gan. Zhong-Yuan, M. C. Tarczynski, B. Shen, 2008 A phenylalanine in DGAT is a key determinant of oil content and composition in maize Nat Genet, 40 3 367 372

107. C. Zhu, S. Naqvi, T. Capell, P. Christou, 2009 Metabolic engineering of ketocarotenoid biosynthesis in higher plants Archives of Biochemistry and Biophysics, 483 2 182 190

Chapter 3

Evaluation of Maize Germplasm for Resistance to Aflatoxin Accumulation

W. Brien Henry[1], Gary L. Windham[1] and Michael H. Blanco[2]

[1]USDA-ARS, Corn Plant Host Resistance Research Unit, Box 9555, Mississippi State, MS 39762, USA;

[2]GEM Coordinator, USDA-ARS, Plant Introduction Research Unit, G212 Agronomy Hall, Iowa State University, Ames, IA 50011, USA; Email: michael.blanco@ars.usda.gov

ABSTRACT

Aflatoxin contamination of maize grain threatens human food and animal feed safety. Breeding for reduced grain aflatoxin accumulation is one of the best strategies presently available to lower grain aflatoxin accumulation. Previously identified sources of germplasm with reduced grain aflatoxin accumulation are excessively tall and late maturing. The objective of this research was to screen germplasm and identify potential sources of aflatoxin resistance. KO679Y and CUBA117:S15-101-001-B-B-B-B inbreds were evaluated for aflatoxin accumulation alongside resistant and susceptible checks with both performing well. These two lines were also evaluated in various crosses. KO679Y performed especially well in crosses with Mp494 and Mp717, resulting in low ear rot and very low aflatoxin levels, but not well in other crosses. A breeding cross including CUBA117:S15-101-001-B-B-B-B as a parent accumulated low levels of aflatoxin both years it was evaluated. Lines resulting from these crosses are being advanced for further evaluation and improvement. KO679Y and CUBA117:S15-101-001-B-B-B-B may prove useful for breeders seeking germplasm sources for ear rot and mycotoxin reduction, especially KO679Y which matures a week earlier and is approximately 25% shorter than current lines resistant to grain aflatoxin accumulation.

INTRODUCTION

Aspergillus flavus produces aflatoxin which is one of the most problematic mycotoxins [1,2]. Mycotoxin contamination of maize grain threatens both human food and animal feed safety [3,4,5,6]. If aflatoxin levels exceed legal limits, farmers in the United States may be unable to market their grain thereby suffering economic losses. Humans with limited food options, such as those of subsistence farmers in developing countries, often must consume aflatoxin-contaminated grain resulting in an increased incidence of liver cancer. Hepatocelluar carcinoma occurs up to 32 times more frequently in developing countries than in developed countries [7]. Chronic exposure of humans to aflatoxin in food results in suppressed growth rates of children and decreased life expectancies. Acute exposure can result in toxicosis and rapid death of animals that consume toxin laden grain.

Breeding maize (*Zea mays* L.) for resistance to *A. flavus* infection and subsequent grain aflatoxin accumulation is considered to be one of the best and most cost effective strategies currently available to reduce aflatoxin losses [8]. Since the early 1970s breeding efforts to lower grain aflatoxin accumulation have resulted in the release of four Mississippi lines Mp313E, Mp420, Mp715 and Mp717 [9,10,11]. Two recently released lines, Mp718 and Mp719 were derived from an Mp715 by Va35 cross resulting in resistant material that is 13 days earlier than the resistance source, Mp715 [12]. Additional maize lines that limit aflatoxin accumulation have been released by researchers in the Southeast United States. The USDA-ARS at Tifton, GA released GT603 with improved agronomic characteristics compared with other resistance sources previously released [13]. Texas A&M researchers released three lines Tx736, Tx739 and Tx740 with both favorable agronomic and reduced toxin accumulation compared with commercial lines [14]. Additional sources of resistance are currently being sought using conventional breeding, marker assisted breeding, and association mapping to identify and develop lines by USDA-ARS Mississippi State, MS. Preliminary marker studies suggest that resistance to grain aflatoxin accumulation is a polygenic trait with multiple quantitative trait loci (QTL) of varying significance within a given line [15]. QTL's also differ among location and significance across resistant inbred lines suggesting the existence of multiple mechanisms for imparting reduced grain aflatoxin accumulation.

Evaluating new germplasm in replicated field trials with known resistant and susceptible checks can assist in the identification of new, and ideally robust, sources of mycotoxin resistance. This technique has been used to screen potential new sources of resistance provided by the Germplasm Enhancement of Maize (GEM) Project. The GEM Project is a collaborative effort of USDA-ARS, universities, private industry, international agricultural research centers and non-governmental organizations (NGO) to broaden the germplasm base of maize [16,17]. The GEM Project's goal is the development of adapted germplasm for the US Corn Belt utilizing tropical and temperate exotic germplasm. Li *et al.* [18] reported that substantial variability existed among maize breeding crosses for aflatoxin resistance from germplasm evaluated from the GEM Project. It was suggested that the most resistant breeding crosses be used to develop new inbred lines. The objectives of this study were (1) to evaluate aflatoxin accumulation in inbred lines and

(2) to evaluate aflatoxin accumulation in breeding crosses and single-cross hybrids constructed with potential sources of resistance.

RESULTS AND DISCUSSION

Maize inbred lines, hybrids and crosses were screened for reduced aflatoxin accumulation in 2009–2011. In 2009, inbred line KO679Y was numerically the third lowest of the 30 lines tested, and statistically the same as the four resistant checks, entries 21, 22, 25, and 26 (Table 1).

Table 1: Aflatoxin evaluation of corn inbreds in 2009

Table 1. Aflatoxin evaluation of corn inbreds in 2009.						
Entry	Pedigree	GEM Code	Aflatoxin[†]	DTM[§]		
			ng g^{-1}			
16	FS8B(T):N11a08c-030-002-B-B	GEMN-0179	7103	a[‡]	62.7	ij
28	NEI9008:S17c21-091-001-B-B	GEM EXP	2629	ab	64.7	g–j
24	GA209	.	2556	ab	72.3	b
27	BR51721:N2012-164-002-B-B-SIB	GEMN-0157	2482	ab	65.7	f–j
23	SC212M	.	2476	ab	73.0	b
2	2132-03_DK888_S11_F2S4_9187-Blk22/00-Sib-B-B	GEMS-0016	2255	ab	64.7	g–j
20	BR52060:S0210-147-001-B-B-B	GEMS-0143	2169	ab	66.0	e–i
19	UR11003:S0302-1011-001-B-B-B	GEMS-0118	2086	a–c	63.3	h–j
4	BR51675:N0620-033-001-B-B-B	GEMN-0140	2063	a–d	64.0	h–j
14	1895-001/98_DKXL370AN11F2S3_7521-29-B-B	GEMN-0133	1660	a–e	68.0	c–g
30	UR11003:S0302-937-001-B-B-Sib	GEMS-0100	1595	a–e	62.0	j
5	AR16026:S17-010-001	GEMS-0061	1272	b–f	69.3	b–f
17	DK888:N11a08a-395-001-B-B	GEMN-0177	980	b–g	69.7	b–e

15	CHIS740:S1411a-783-002-B-B-Sib	GEMS-0091	859	b–g	62.7	ij
29	CH05015:N15-003-001-B-B-SIB-SIB	GEMN-0071	654	b–h	67.0	d–h
12	2258-03_XL380_S11_F2S4_71/97_Bulk/98-Sib-B-B-B-B	GEMS-0030	596	b–h	70.3	b–d
1	CUBA164:S1517-163-001-B-B	GEMS-0074	440	c–i	64.7	g–j
3	2088-01_DK212T_S11_F2S4_9157-Blk29/00-sib-B-B-B	GEMS-0006	425	d–j	68.0	c–g
7	1881-002/98_DKXL370AN11F2S3_7521-05-B-B	GEMS-0128	405	e–j	67.0	d–h
8	1886-003/98_DKXL370AN11F2S3_7521-05-B	GEMN-0132	319	f–k	71.0	bc
6	FS8A(S):S09-43-2	GEMS-0002	318	f–k	62.7	ij
10	1883-002/98_DKXL370AN11F2S3_7521-05-B-B	GEMN-0131	232	g–l	69.7	b–e
22	Mp313E	.	221	g–m	79.5	a
13	CUBA117:S15-101-001-B-B-B-B	GEM EXP	215	h–m	79.0	a
21	Mp717	.	150	h–m	77.0	a
9	1883-001/98_DKXL370AN11F2S3_7521-05-B-B	GEMN-0130	104	i–m	70.3	b–d
26	Mp494	.	87	j–m	80.0	a
18	KO679Y	PI 591017	75	k–m	70.3	b–d
11	DKXL380:N11a18-059-001	GEM EXP	58	l–m	68.0	c–g
25	Mp317	.	45	m	79.0	a

† Means for aflatoxin concentration were transformed [ln (y + 1)] before statistical analysis, and tests for significance were performed on transformed means prior to conversion back to the original scale; ‡ Means in a column followed by the same letter do not differ at *P*

Of the four low-aflatoxin accumulating checks that were included in the trial, KO679Y numerically accumulated lower toxin, 75 ng g^{-1}, than three of the entries and was not statistically different from any of them. KO679Y was included in the 2010 inbred aflatoxin evaluation and had the 6th lowest aflatoxin level of the 20 entries, 223 ng g^{-1} (Table 2).

Table 2: Aflatoxin evaluation of corn inbreds in 2010

Entry	Pedigree	GEM Code	Aflatoxin † ng g^{-1}	DTM §		
17	Mp317				85.0	a
20	SC212M		5247	a ‡	81.3	b
10	DKXL380:S08a12-288-001-B-B	GEM EXP	3294	ab	75.0	c–e
19	GA209	.	2155	a–c	78.0	c
8	DKXL380:S08a12-069-001-B-B	GEM EXP	1963	a–c	73.7	ef
9	DKXL380:S08a12-165-001-B-B	GEM EXP	1422	a–d	71.0	f
7	DKXL380:N11a18-059-001-B-B	GEM EXP	1243	a–e	73.3	ef
13	FS8B(T):N11a-087-001-B-B-SIB-B	GEMN-0097	1092	a–e	71.0	f
15	DKB844:S1601-073-001-B-B-B-B-B	GEMS-0115	817	a–e	76.0	c–e
3	DKXL380:S11-B-004-003-B-B-Sib-B-B-B-B-B	GEMS-0030	653	b–e	76.0	c–e
5	AR16026:S17-010-001-B-B-B	GEMS-0061	539	b–e	75.0	c–e
16	Ki21	.	401	c–e	81.3	b
6	DKXL370A:N11-B-005-010-B-B	GEMN-0130	333	c–e	76.0	c–e
11	DKXL380:S08a12-294-001-B-B	GEM EXP	254	d–e	74.7	de
14	KO679Y	PI 591017	223	de	77.0	cd
12	CUBA117:S15-101-001-B-B-B-B-B-B-B	GEM EXP	209	ef	73.3	ef
2	DKXL370A:S11-B-025-001-B-B-Sib-B-B	GEMS-0027	207	ef	71.0	f
18	Mp717	.	33	fg	82.0	ab
4	DKXL380:S11-B-029-001-B-B-Sib-B-B-B	GEMS-0032	28	g	83.0	ab
1	DKXL370A:S11-B-025-002-B-B-Sib	GEMS-0028	24	g	73.7	ef

Another GEM experimental line from Cuba, CUBA117:S15-101-001-B-B-B-B, fared equally well with 215 and 209 ng g^{-1} aflatoxin accumulation in 2009 and 2010, respectively. This line was statistically similar to all the resistant checks both years (Table 1, Table 2).

Four of the entries in the 2010 inbred evaluation (DKXL entries 1–4) were identified as low grain aflatoxin accumulators in prior evaluations [19], and for the most part they performed well in 2010 (Table 2). Mp317 and Mp717 were included as resistant checks, but Mp317 did not produce ears suitable for inoculation or harvest because of the heat and drought.

In addition to favorable toxin data, KO679Y was significantly earlier than the resistant checks included in the inbred evaluations. Compared with the resistant checks, KO679Y was approximately 10 days earlier in 2009 and 7 days earlier in 2010 (Table 1, Table 2).

Approximately 150 breeding crosses and 50 hybrid crosses were evaluated in 2008 for aflatoxin accumulation (data not shown), and several of the best entries were evaluated for aflatoxin accumulation again in 2009 (Table 3).

In addition to repeated evaluation of some of the better performing crosses from 2008, additional breeding crosses containing DKXL or CUBA117:S15-101-001-B-B-B-B germplasm were also included in the 2009 trial (Table 3). Of the breeding crosses identified with low aflatoxin accumulation in 2008 and retested in 2009, entries 32 through 36 fared well with toxin levels at or below 100 ng g^{-1}. Conversely, entries 31, 37, and 38 were all above 300 ng g^{-1} (Table 3).

Table 3: Aflatoxin evaluation of corn hybrids and breeding crosses in 2009 - See more at: http://www.mdpi.com/2073-4395/2/1/28/htm#sthash.jfR7W-wm4.dpuf

Entry	Pedigree/GEM code	Aflatoxin † ng g^{-1}		
45	CML323:N1550	1556	a ‡	
47	NEI9004:N0803	1043	ab	
16	GEMS-0073/GEMN-0132	829	a–c	
13	GEMS-0016/GEMN-0140	665	a–d	
30	GEMN-0095/LH200	608	a–d	
40	(GEMS-0002/GEMS-0003)-B	590	a–e	
38	CHIS462:S0405	556	a–e	
31	CHIS462:S04	528	a–e	

44	GUAT209:N11c50	503	a–f	
10	GEMS-0150/GEMN/0140	488	a–f	
29	GEMN-0094/LH200	473	a–g	
8	GEMS-0116/GEMN-0132	459	a–h	
42	NS1/GEMN-0133	441	a–i	
46	TZISTRI112/PHB47	392	a–j	
20	GEMN-0133/HC33	365	a–j	
7	GEMS-0016/GEMN-0097	361	a–j	
1	DKXL380:N11a18-059-001/LH200	340	a–k	
37	FS8B(S):S17a	334	a–k	
19	GEMS-0091/GEMN-0133	317	a–k	
39	(CUBA164:S2012-459-001-B/GEMS-0002)-B	317	a–k	
11	GEMS-0162/GEMN-0133	308	a–k	
9	GEMS-0115/GEMN-0133	305	a–l	
15	GEMS-0061/GEMN-0179	305	a–l	
2	DKXL380:N11a18-420-001/LH200	284	a–l	
12	GEMS-0116/GEMN-0133	262	a–l	
26	GEMS-0175/GEMN-0133	262	a–l	
4	DKXL380:N11a18-059-001/HC33	250	b–l	
14	GEMS-0016/GEMN-0177	224	b–l	
25	GEMS-0115/GEMN-0132	161	c–m	
3	GEMS-0074/LH185	153	c–n	
5	GEMS-0118/GEMN-0140	151	c–n	
43	(S21z/GEMS-0115)-B	148	c–n	
17	GEMS-0091/GEMN-0132	112	d–n	
35	BR106:T33a/LH132	100	e–n	
50	T173 x Mp717	92	f–n	
34	CUBA164/NEI9004	91	f–n	
18	GEMS-0031/GEMN-0133	79	g–o	
36	BR105/NEI9004	79	g–o	
28	GEMS-0061/GEMN-0130	77	h–o	
27	GEMS-0182/GEMN-0132	74	i–o	
21	GEMS-0031/GEMN-0131	72	j–o	
33	(UR13085/N28)/Cuba117:S1520-156-001-B	72	j–o	
6	GEMS-0143/GEMN-0140	65	j–o	
24	CUBA117:S15-101-001-B-B-B-B-B/GEMN/0130	59	k–o	
32	(UR13085/N28)/GEMS-0002	57	k–o	
41	Ki 14:S21z	51	l–o	

23	CUBA117:S15-101-001-B-B-B-B-B/GEMN-0140	28	m–o	
22	GEMS-0031/GEMN-0132	26	no	
48	Mp313E × Mp717	26	no	
49	Mp313E × Mo18w	15	o	

‡ Means in a column followed by the same letter do not differ at *P*

The 2010 evaluation of crosses included four entries with KO679Y in the pedigree (Table 4).

Of the four entries, entry 9, accumulated 163 ng g^{-1} of aflatoxin which was the 7th lowest entry, numerically; however, none were statistically different than the susceptible check. Two entries, 17 and 18, included Mp717 as a parent and generated very low aflatoxin levels with 5 and 34 ng g^{-1}, respectively. This was expected because crosses resulting from Mp717 as a parent usually have lowered grain aflatoxin accumulation. These two entries were included because they are known to have resistance (Table 3). On the contrary, entries 8 and 4 both included Mp717 as a parent, albeit in a lesser percentage, but the grain aflatoxin accumulation was 253 and 430 ng g^{-1}, respectively. It appears that the susceptibility of the other portions of the pedigree may have overwhelmed the resistance contributions of Mp717. This may also have been the case with entry 5, and perhaps with the other entries with KO679Y in this particular evaluation; however, these explanations are speculative and would require additional testing to confirm. KO679Y did not appear to combine well for lowered grain aflatoxin accumulation in any of the four combinations evaluated in 2010 (Table 3).

Entry 10, a breeding cross containing the CUBA117:S15-101-001-B germplasm, performed consistently well in both 2009 and 2010 with 28 and 33 ng g^{-1}, respectively.

Table 4: Aflatoxin evaluation of corn hybrids and breeding crosses in 2010

Entry	Pedigree	Aflatoxin † ng g^{-1}	
1	(KO679Y/GEMS-0115)-B	573	a ‡
19	SC212M × GA209	515	a
20	T173 × SC212M	487	a

5	((KO679Y/GEMS-0115)/GEMS-0162)	454	a
3	(CUBA117:S15-101-001-B-B-B-B-B/GEMS-0027)-B	433	a
4	((GEMS-0115/Mp717)/GEMS-0162)	430	a
2	(GEMN-0140/GEMN-0130)-B	319	a
15	(UR13085/N28)/Cuba117:S1520-156-001-B-B-B (71-Days, Late)	302	a
12	KO679Y/GEMN-0097	276	ab
8	((GEMS-0115/Mp717)/Ki21)	253	ab
16	Mp317 × Jala-B-B	215	ab
11	CUBA117:S15-101-001-B-B-B-B-B/GEMN-0130	196	ab
6	((87916W/Mp717)/4676A)	174	ab
9	((KO679Y/GEMS-0115)/GEMS-0181)	163	ab
7	((GEMS-0115/Mp717)/GEMS-0061)	144	ab
14	UR13085:S99g99u-B-B-B (61 days, Early)	138	ab
13	CUBA117:S15-101-001-B-B-B-B/GEMN-0130	92	ab
18	Mp317 × Mp717	34	bc
10	CUBA117:S15-101-001-B-B-B-B-B/ GEMN-0140	33	bc
17	Mp313E × Mp717	5	c

Means for aflatoxin concentration were transformed [ln (y + 1)] before statistical analysis, and tests for significance were performed on transformed means prior to conversion back to the original scale; ‡ Means in a column followed by the same letter do not differ at P - See more at: http://www.mdpi.com/2073-4395/2/1/28/htm#sthash.jfR7Wwm4.dpuf

The 2011 evaluation of crosses had a wide range of aflatoxin accumulation ranging from 0 to 776 ng g^{-1}(Table 5).

Two of the top three entries included KO679Y as a parent. Entry 7, Mp717 × KO679Y, and entry 8, KO679Y × Mp494, accumulated very low levels of aflatoxin at 1 and 2 ng g^{-1}, respectively. Each of these two entries also had very low ear rot scores (Table 4). Ear rot was scored from 0 to 9 with 0 representing no ear rot and 9 representing an ear fully covered with rot. The mean score for this evaluation was 3.4 and the scores for Mp717 × KO679Y and KO679Y × Mp494 were 0.3 and 1.0, respectively. KO679Y did not reduce ear rot across all of the single-cross hybrid combinations. Entry 1, GEMS-0030 × KO679Y, was the second most susceptible entry with an ear rot score of 6.3 (Table 4). It is of interest that entry 8, the KO679Y × Mp494 cross, had low ear rot scores because, although Mp494 is known for low grain aflatoxin accumulation, it often has significant ear rot both as an inbred and in crosses [20]. The KO679Y × Mp494 cross appears to have overcome the ear rot deficiencies of Mp494, while also maintaining reduced grain aflatoxin accumulation.

Table 5: Aflatoxin evaluation of corn hybrids and breeding crosses in 2011

Entry	**Pedigree**	**Aflatoxin** † **ng g^{-1}**	**Ear Rot**§		
18	GA209 × SC212M	776	a ‡	4.0	c–f
12	(GEMS-0115/Mp717)/TZAR101	683	a	5.7	a–d
13	(GEMS-0115/MP717)/GEMS-0115	386	ab	6.0	a–c
17	GEMS-0100 × SC212M	381	ab	4.0	c–f
15	KUI44:N99ap	290	ab	7.3	a
14	KO679Y/(GEMS-0002/GEMS-0003)	174	a-c	4.0	c–f
2	NEI9008:S17c21-091-001-B-B-B × KO679Y	112	a-d	2.0	f–h
3	GEMS-0115 × KO679Y	96	a-d	3.7	d–f
5	KO679Y × Mp317	69	b-e	3.0	e–g
4	CUBA117: S15-101-001-B-B-B-B × KO679Y	62	b-e	3.3	ef
1	GEMS-0030 × KO679Y	58	b-e	6.3	ab
6	KO679Y × Mp313E	53	b-e	4.3	b–d
16	Mp313E × GEMN-0157	28	c-e	2.0	f–h
10	Mp317 × GEMS-0115	24	c-e	3.0	e–g
19	Mp494 × NEI9008:S17c21-091-001-B-B-B	15	d-f	0.3	h
9	Mp317 × NEI9008:S17c21-091-001-B-B-B	13	d-f	4.7	b–d
11	GEMS-0030 × Mp317	9	ef	1.0	gh
8	KO679Y × Mp494	2	fg	1.0	gh
7	Mp717 × KO679Y	1	fg	0.3	h
20	Mp717 × Mp313E	0	g	1.0	gh

‡Means in a column followed by the same letter do not differ at *P*

Entry 4 was comprised of two of the more promising potential sources of resistance, inbred lines CUBA117: S15-101-001-B-B-B-B and KO679Y; however, it accumulated 62 ng g^{-1}, a value that suggests only moderate resistance, lower than the susceptible checks, but higher than the best resistant crosses and resistant checks.

Both Mp717 and Mp494 were previously identified resistance sources for reduced grain aflatoxin accumulation and each of these inbred lines has a hard, flinty kernel type that appears to have responded well with respect to ear rot and grain aflatoxin accumulation in crosses with KO679Y. Resistance sources Mp317 and Mp494 also combined well with a GEM line having Thai germplasm in 25% of the pedigree, NEI9008: S17c21-091-001-B-B-B, as entries 19 and 9. These crosses were statistically similar to the two best crosses that included KO679Y as a parent. It is interesting that *per se* the GEM line, NEI9008: S17c21-091-001-B-B-B generated elevated grain aflatoxin levels statistically similar to, and numerically higher than, the susceptible checks GA209 and SC212M (Table 1); however, if crossed to resistant lines the resulting crosses perform very well.

EXPERIMENTAL SECTION

Planting Dates and Experiments

Multiple experiments to evaluate inbreds, crosses and hybrids for aflatoxin accumulation were conducted at the Rodney Foil Research Facility at Mississippi State University (MSU). The 2009 inbred evaluation was planted on 29 April; the breeding cross evaluation and a second inbred evaluation were planted 14 April, 2010, and the single-cross hybrid evaluation was planted on 10 May 2011. The soil type for the 2009 and 2010 experiments was a Marietta fine sandy loam and for the 2011 experiment was a Leeper silty clay loam.

Germplasm

Seed for the initial evaluations were received from the GEM Project in 2009. One of the lines that performed well in these trials is KO679Y (PI 591017), a public South African inbred obtained from the GEM Project located at the North Central Regional Plant Introduction Station (NCRPIS) in Ames, IA. This inbred has been noted for its excellent ear quality and has been used by the GEM Project as a potential source of resistance to Fusarium and Diplodia ear rot diseases. It is one among many potential sources of resistance evaluated at the USDA-ARS location at Mississippi State. GEMS-0002 was described as a registered release in 2005 [21], and GEMS-0006, GEMS-0028, and GEMS-0030 were registered in 2006 [19]. Nursery seed increase with agronomic evaluation and inbred aflatoxin screening were conducted that year. Favorable toxin data in 2009 resulted in additional evaluations in 2010, and a request to GEM for additional material to evaluate. Material was received from the GEM Project in 2010 and evaluated alongside other experimental crosses including single-cross resistant and susceptible checks. Based on previous results, KO679Y was used as a parent of 8 additional single cross hybrids which were then evaluated alongside other experimental crosses including single-cross positive and negative checks. Single crosses were made at Mississippi State, MS in 2010 from paired row crosses. Four of these inbreds were older Southern lines evaluated previously for aflatoxin accumulation both as inbreds and in various single-cross hybrid combinations. These lines, Mp717, Mp313E, Mp317 and Mp494, previously exhibited resistance to aflatoxin accumulation [19,22,23]. Across most environments, these four inbred lines are generally tall and late-maturing, and are used as sources of resistance to grain aflatoxin accumulation. The other four inbred lines crossed to KO679Y were selections from the GEM Project. Two of these named inbred entries, GEMS-0030 (low aflatoxin) and GEMS-0115 (good yield), were released previously through GEM in 2003 and 2005 respectively. The other two lines to which KO679Y was crossed were CUBA117: S15-101-001-B-B-B-B and NEI9008: S17c21-091-001-B-B-B which are both experimental GEM lines. The Cuban line exhibited somewhat reduced grain aflatoxin accumulation over multiple years and the NEI9008 (Thai) line did not exhibit resistance *per se*; however, in another diallel experiment (data not shown) and in the 2011 crosses evaluation (Table 4) it combined well with older Southern resistant lines to generate single-cross hybrids with low grain aflatoxin accumulation. These two lines also were provided by the GEM Project. The GEM lines other than KO679Y were derived from self-pollination in accordance with GEM protocol using the pedigree breeding method. Phenotypic traits and protocols for breeding and writing pedigrees can be found on the GEM Project website [24].

Plot size and Management

Entries in the inbred, breeding cross, and single-crosshybrid evaluations at MSU were planted in 5.1 m long single-row plots, with rows spaced 0.97 m apart, and thinned to 20 plants per row in a randomized complete block with three replicates. Fertilizer and herbicides were applied in accordance with standard local production practices and at no time throughout the season was weed pressure or fertility limiting for crop growth. All years experienced periodic drought conditions and supplemental furrow irrigation was applied once in 2010 and multiple times in 2009 and 2011, to assure that there would be enough grain to harvest and evaluate.

Fungal Inoculum and Aflatoxin

A. flavus isolate NRRL 3357 has produced high levels of aflatoxin in maize grain in prior studies [25,26,27] and was used to increase the inoculum in these experiments. Inoculum was increased according to Windham *et al.* [26]. Briefly, *A. flavus* was grown on sterile maize cob grits (size 2040, Grit-O-Cobs®, The Andersons, Maumee, OH 43537, USA) in 500-ml flasks. Each contained 50 g of grits and 100 mL of sterile, distilled water, and was incubated at 28 °C for 3 weeks. Conidia in each flask were washed from the grits using 500 mL sterile distilled water containing 0.1% Tween 20 per liter and filtered through four layers of sterile cheesecloth. A tree marking gun was used to deliver a 3.4-mL spore suspension containing 3×10^6 *A. flavus* conidia at 7 days after midsilk. The solution was injected underneath the husks near the middle of the ear.

Ears were harvested approximately 60 days following inoculation and placed in a forced air drier at 38 °C for 5 to 7 days. Following drying, the ears from each plot were bulked together, shelled, mixed thoroughly and ground with a Romer Mill (Union, MO). Aflatoxin concentration was determined with the Vicam Aflatest (Watertown, MA) which detects aflatoxin levels as low as 1 ng g^{-1}.

Statistical Analysis

All statistical analyses were conducted using the SAS software package (version 8.2; SAS Institute Inc., Cary, NC, USA). Logarithmic transformation [log (y + 1)] was used on all aflatoxin data to stabilize the variance. Data were analyzed with the PROC GLM procedure and means were separated using Fisher's protected least significant difference test at P = 0.05. Aflatoxin data are reported as geometric means (antilog of the logarithmic mean).

CONCLUSIONS

In summary, results show that inbred lines KO679Y and Cuba117:S15-101-001-B-B-B-B are promising sources of reduced grain aflatoxin accumulation that should be further evaluated in replicated field trials over multiple years. CUBA117:S15-101-001-B-B-B-B in one of the breeding cross combinations produced very low grain aflatoxin accumulation in 2009 and 2010. This line is under further development in our breeding program. Because there is a shortage of breeding material for reduced grain aflatoxin accumulation, it is helpful to plant breeders to identify and document new sources of germplasm. This is particularly true throughout the southern US, and tropical regions such as Africa where aflatoxin is a serious problem. In addition to being earlier than presently available sources of aflatoxin resistance, KO679Y is also 25% shorter than other currently available inbred lines. These favorable agronomic characteristics increase the likelihood of KO679Y being used in future breeding programs. Inbred toxin accumulation values for KO679Y compared to known checks were very low in 2009 and moderately low in 2010. Crosses including both KO679Y and CUBA117:S15-101-001-B-B-B-B as parents varied in aflatoxin accumulation. KO679Y appears to combine well with Mp717 and Mp494 (for reduced aflatoxin) both of which are late and tall and have hard, flinty, yellow-orange kernel types. Advancement of the KO679Y × Mp494 cross is underway in our winter nursery. Should these and other KO679Y crosses perform well in upcoming evaluations, it would be worth considering generating a mapping population and identifying specific QTL in KO679Y associated with reduced grain aflatoxin accumulation.

ACKNOWLEDGEMENTS

The authors thank the GEM Project for providing some of the genetic resources upon which this study was based. The authors also thank

USDA-ARS technicians Alan Hames, Mike Alpe, Paul Buckley, Greg Flint, LaDonna Owens, Jack Haynes, and Boo Mathews, for their assistance with toxin preparation and analyses, field management, and harvest. Mention of trade names or commercial products in this manuscript is solely for the purpose of providing scientific information and does not imply recommendation or endorsement by the USDA.

REFERENCES

1. McGregor, D.; Castegarno, M. Carcinogenic risk assessment of mycotoxins. *Rev. Med. Vet.* 1998, *149*, 671–678.
2. Payne, G. Aflatoxin in maize. *Crit. Rev. Plant Sci.* 1992, *10*, 423–440.
3. Schepers, J.S.; Balazs, E. The mycotoxin threat to world safety. *Int. J. Food Micro.* 2007, *119*, 1–2.
4. Richard, J.; Handl, J.; Chin, L.J.; Tan, L.M.; Binder, E.M. Worldwide occurrence of mycotoxins in commodities, feeds and feed ingredients. *Anim. Feed Sci. Technol.* 2007, *137*, 265–282.
5. Wu, F. Measuring the economic impacts of *Fusarium* toxins in animal feeds. *Anim. Feed Sci. Technol.* 2007, *137*, 363–374.
6. Wu, F. Mycotoxin reduction in Bt corn: potential economic, health, and regulatory impacts. *Transgenic Res.* 2006, *15*, 277–289.
7. Wu, F.; Lui, Y. Global burden of aflatoxin-induced hepatocellular carcinoma: A risk assessment. *Environ. Health Perspect.* 2010, *118*, 818–824.
8. Williams, W.P. Breeding for resistance to aflatoxin accumulation in maize. *Mycotoxin Res.* 2006, *22*, 27–32.
9. Zummo, N.; Scott, G. Registration of Mp313E parental line of maize. *Crop Sci.* 1990, *31*, 1374–1375.
10. Windham, G.L.; Williams, W.P. Registration of dent corn germplasm line Mp715. *Crop Sci.* 2001, *41*, 1374–1375.
11. Windham, G.L.; Williams, W.P. Diallel analysis of fumonisin accumulation in maize. *Field Crops Res.* 2009, *114*, 324–326.
12. Windham, G.L.; Williams, W.P. Registration of Mp718 and Mp719 germplasm lines of maize. *J. Plant Regis.* 2012, *6*, 1–3.

13. Widstrom, N.W.; McCoy, A.E.; Lee, R.D.; Scully, B.T.; Ni, X.; Krakowsky, M.D.; Guo, B.Z. Registration of inbred line GT603. *J. Plant Regis.* 2011, *5*, 211–214.
14. Landivar, J.C.; Tooney, W.L.; Murray, S.C.; Odvody, G.; Isakeit, T.; Betran, F.J.; Mayfield, K. Registration of maize germplasm lines Tx736, Tx739, and Tx740 for reducing preharvest aflatoxin accumulation. *J. Plant Regis.* 2012, *6*, 1–7.
15. Williams, W.P.; Windham, G.L.; Shan, X.; Krakowsky, M.D.; Brooks, T.D.; Warburton, M.L. Identification and mapping of new sources of resistance to aflatoxin accumulation in maize. *Crop Sci.* 2009, *49*, 1403–1408.
16. Pollak, L.M. The history and success of the public-private project on germplasm enhancement of maize GEM. *Adv. Agron.* 2003, *78*, 45–87.
17. Blanco, M.; Trevisan, W. *The Importance of the Germplasm Enhancement of Maize Project (GEM) for US Seed Companies*, 44th ed.; Annual Illinois Corn Breeding School: Illinois, IL, USA, 2008.
18. Pollak, L.M.; Moreno, O.J.; Kang, M.S.; Li, R. Field resistance to *Aspergillus flavus* from exotic maize (*Zea mays* L.) germplasm..
19. Goodman, M.; Carson, M.L.; Holley, R.; Clements, M.; Holland, J.; Duvick, S.; Millard, M.; Blanco, M.; Balint-Kurti, P. Registration of 20 maize breeding germplasm lines adapted to the southern USA. *Crop Sci.* 2006, *46*, 996–998.
20. Hawkins, L.K.; Windham, G.L.; Williams, W.P.; Henry, W.B. Evaluation of maize inbred lines for resistance to Aspergillus and Fusarium ear rot and mycotoxin accumulation. *Agron. J.* 2009, *101*, 1219–1226.
21. Montgomery, K.T.; Pollak, L.M.; Pratt, R.C. Registration of maize germplasm line GEMS-0002. *Crop Sci.* 2005, *45*, 2130–2131.
22. Windham, G.L.; Williams, W.P.; Gardner, H.G. Diallel analysis of aflatoxin accumulation in maize. *Field Crops Res.* 2007, *102*, 60–63.
23. Buckley, P.M.; Windham, G.L.; Williams, W.P. Diallel analysis of aflatoxin accumulation in maize. *Crop Sci.* 2008, *48*, 134–138.
24. *GEM Project Website*. Available online: www.public.iastate.edu/usda-gem (accessed on 1 Feb 2012).

25. Williams, W.P.; Windham, G.L. *Aspergillus flavus* infection and aflatoxin accumulation in resistant and susceptible maize hybrids. *Plant Dis.* 1998, *82*, 281–284.
26. Williams, W.P.; Windham, G.L. Evaluation of corn inbred lines and advanced breeding lines for resistance to aflatoxin contamination in the field. *Plant Dis.* 2002, *86*, 232–234.
27. Brooks, T.D.; Hawkins, L.K.; Williams, W.P.; Windham, G.L. Effect of *Aspergillus flavus* inoculation methods and environmental conditions on aflatoxin accumulation in corn hybrids. *J. Toxicol. Toxin Rev.* 2009, *28*, 70–78.

Chapter 4

Development of Maize Host Resistance to Aflatoxigenic Fungi

Robert L. Brown[1], Deepak Bhatnagar[1], Thomas E. Cleveland[1], Zhi-Yuan Chen[2] and Abebe Menkir[3]

[1] USDA-ARS, Southern Regional Research Center, New Orleans, LA, USA

[2] Department of Plant Pathology and Crop Physiology, Louisiana State University Agricultural Center, Baton Rouge, LA, USA

[3] International Institute of Tropical Agriculture, Ibadan, Nigeria

INTRODUCTION

Aflatoxins, the toxic and highly carcinogenic secondary metabolites of *Aspergillus flavus* and *A.parasiticus* are the most widely investigated of all mycotoxins because of their central role in establishing the significance of mycotoxins in animal diseases, and the regulation of their presence in food [1, 2]. Aflatoxins pose serious health hazards to humans and domestic animals, because they frequently contaminate agricultural commodities [3, 4]. Presently, numerous countries have established or proposed regulations for controlling aflatoxins in food and feeds [5]; the US Food and Drug Administration (FDA) has limits of 20 ppb, total aflatoxins, on interstate commerce of food and feed, and 0.5 ppb of aflatoxin M1 on the sale of milk. However, many countries, especially in the developing world, experience contamination of domestic-grown commodities at alarmingly greater levels than does the U.S. Evidence of this was shown in a study that revealed a strong association between exposure to aflatoxin and both stunting (a reflection of chronic malnutrition) and being underweight (a reflection of acute malnutrition) in West African children [6]. Also, a 2004 outbreak of acute aflatoxicosis in Kenya, due to the ingestion of contaminated maize, resulted in 125 deaths [7].

Recognition of the need to control aflatoxin contamination of food and feed grains has elicited responses outlining various approaches from researchers to eliminate aflatoxins from maize and other susceptible crops. The approach to enhance host resistance through breeding gained renewed attention following the discovery of natural resistance to *A. flavus* infection and aflatoxin production in Maize [8-12]. While several resistant maize genotypes have been identified through field screening, there is always a need to continually identify and utilize additional sources of maize genotypes with aflatoxin-resistance.

An important contribution to the identification/investigation of kernel aflatoxin-resistance has been the development of a rapid laboratory screening assay. The kernel screening assay (KSA), was developed and used to study resistance to aflatoxin production in GT-MAS:gk kernels [13, 14]. The KSA is designed to address the fact that aflatoxin buildup occurs in mature and not developing kernels. Although, other agronomic factors (e.g. husk tightness) are known to affect genetic resistance to aflatoxin accumulation in the field, the

KSA measures seed-based genetic resistance. The seed, of course, is the primary target of aflatoxigenic fungi, and is the edible portion of the crop. Therefore, seed-based resistance represents the core objective of maize host resistance. Towards this aim, the KSA has demonstrated proficiency in separating susceptible from resistant seed [13, 14]. This assay has several advantages, as compared to traditional field screening techniques [14]: 1) it can be performed and repeated several times throughout the year and outside of the growing season; 2) it requires few kernels; 3) it can detect/identify different kernel resistance mechanisms; 4) it can dispute or confirm field evaluations (identify escapes); and 5) correlations between laboratory findings and inoculations in the field have been demonstrated. The KSA can, therefore, be a valuable complement to standard breeding practices for preliminary evaluation of germplasm. However, field trials are necessary for the final confirmation of resistance.

DISCOVERY OF AFLATOXIN-RESISTANCE

Traditional Screening Techniques

Screening maize for resistance to kernel infection by *Aspergillus flavus* or for resistance to aflatoxin production is a more difficult task than most disease screening. Successful screening in the past had been hindered [15] by the lack of 1) a resistant control; 2) inoculation methods that yield infection/aflatoxin levels high enough to differentiate among genotypes (natural infection is undependable); 3) repeatability across different locations and years; and, 4) rapid and inexpensive methods for assessment of fungal infection and aflatoxin levels. Several inoculation methods, including the pinbar inoculation technique (for inoculating kernels through husks), the silk inoculation technique, and infesting corn ears with insect larvae infected with *A. flavus* conidia have been tried with varying degrees of success [9, 16]. These methods can each be useful, however, clarity must exist as to the actual resistance trait to be measured (e.g. husk tightness; silk traits; the kernel pericarp barrier; wounded kernel resistance), before an appropriate technique can be employed. Silk inoculation, however, (possibly more dependent upon the plant's physiological stage and/or environmental conditions) has proven to be the most inconsistent of the inoculation methods [17].

Plating kernels to determine the frequency of kernel infection and examining kernels for emission of a bright greenish-yellow fluorescence (BGYF) are methods that have been used for assessing *A. flavus* infection [15]. While both methods can indicate the presence of *A. flavus* in seed, neither can provide the kind of accurate quantitative or tissue-localization data useful for effective resistance breeding. Several protocols have been developed and used for separation and relatively accurate quantification of aflatoxins [18].

Early Identification of Resistant Maize Lines

Two resistant inbreds (Mp420 and Mp313E) were discovered and tested in field trials at different locations and released as sources of resistant germplasm [11, 19]. The pinbar inoculation technique was one of the methods employed in the initial trials, and contributed towards the separation of resistant from susceptible lines [11]. Several other inbreds, demonstrating resistance to aflatoxin contamination in Illinois field trials (employing a modified pinbar technique) also were discovered [12]. Another source of resistance discovered was the maize breeding population, GT-MAS:gk. This population was derived from visibly classified segregating kernels, obtained from a single fungus-infected hybrid ear [10]. It tested resistant in trials conducted over a five year period, where a kernel knife inoculation technique was employed.

These discoveries of resistant germplasm may have been facilitated by the use of inoculation techniques capable of repeatedly providing high infection/aflatoxin levels for genotype separation to occur. While these maize lines do not generally possess commercially acceptable agronomic traits, they may be invaluable sources of resistance genes, and as such, provide a basis for the rapid development of host resistance strategies to eliminate aflatoxin contamination.

INVESTIGATIONS OF RESISTANCE MECHANISMS/TRAITS IN MAIZE LINES

Molecular Genetic Investigations of Aflatoxin-Resistant Lines

Chromosome regions associated with resistance to *A. flavus* and inhibition of aflatoxin production in maize have been identified through Restriction Fragment Length Polymorphism (RFLP) analysis in three "resistant" lines (R001, LB31, and Tex6) in an Illinois breeding program, after mapping populations were developed using B73 and/or Mo17 elite inbreds as the "susceptible" parents [20, 21]. Chromosome regions associated with inhibition of aflatoxin in studies considering all 3 resistant lines demonstrated that there are some regions in common. Regions on chromosome arms 2L, 3L, 4S, and 8S may prove promising for improving resistance through marker assisted breeding into commercial lines [21]. In some cases, chromosomal regions were associated with resistance to *Aspergillus* ear rot and not aflatoxin inhibition, and vice versa, whereas others were found to be associated with both traits. This suggests that these two traits may be at least partially under separate genetic control. QTL studies involving other populations have identified chromosome regions associated with low aflatoxin accumulation.

In a study involving 2 populations from Tex6 x B73, conducted in 1996 and 1997, promising QTLs for low aflatoxin were detected in bins 3.05-6, 4.07-8, 5.01-2, 5.05-5, and 10.05-10.07 [22]. Environment strongly influenced detection of QTLs for lower toxin in different years; QTLs for lower aflatoxin were attributed to both parental sources. In a study involving a cross between B73 and resistant inbred Oh516, QTL associated with reduced aflatoxin were identified on chromosomes 2, 3 and 7 (bins 2.01 to 2.03, 2.08, 3.08, and 7.06) [23]. QTLs contributing resistance to aflatoxin accumulation were also identified using a population created by B73 and resistant inbred Mp313E, on chromosome 4 of Mp313E [24]. This confirmed the findings of an earlier study involving Mp313E and susceptible Va35 [25]. Another QTL in this study, which has similar effects to that on chromosome 4, was identified on chromosome 2 [24]. A recent study to identify aflatoxin-resistance QTL and linked markers for marker-assisted breeding was conducted using a population developed from Mp717, an aflatoxin-resistant maize inbred, and NC300, a susceptible inbred adapted to the southern U.S. QTL were identified on all chromosomes, except 4, 6, and 9; individual QTL accounted for up to 11% of phenotypic variance in aflatoxin accumulation [26]. Lastly, in a study of population of F2:3 families developed from resistant Mp715 and a southern-adapted susceptible, T173, QTL with phenotypic effects up to 18.5% were identified in multiple years on chromosomes 1, 3, 5, and 10 [27].

A number of genes corresponding to resistance-associated proteins (RAPs), that were identified in proteomics studies (see section 3.5.1 below) have been mapped to chromosomal location using the genetic sequence of B73 now available online (http://archive.maizesequence.org/index.html) [28]. Using the DNA sequence of the RAPs and blasting them against the B73 sequence allowed us to place each gene into a virtual bin, allowing us to pinpoint the chromosomal location to which each gene maps. The chromosomes involved include the above-mentioned chromosomes 1, 2, 3, 7, 8 and 10, some in bins closely located to those described above. Another study also mapped RAPs to bins on the above-chromosomes as well as chromosomes 4 and 9 [29].

Kernel Pericarp Wax

Kernel pericarp wax of maize breeding population GT-MAS:gk has been associated with resistance to*Aspergillus flavus* infection /aflatoxin production. Previously, kernel wax of GT-MAS:gk was compared to that of 3 susceptible genotypes. Thin layer chromatography (TLC) of wax from these genotypes showed a band unique to GT-MAS:gk and a band unique to the three susceptible lines [30]. GT-MAS:gk kernel wax also was shown to inhibit *A. flavus* growth. A later investigation compared GT-MAS:gk wax resistance-associated traits to that of twelve susceptible maize genotypes [31]. TLC results of wax from these lines confirmed findings of the previous investigation, demonstrating both the unique GT-MAS:gk TLC band and the unique 'susceptible' band. Gas chromatography/mass spectroscopy (GC/MS) analysis of the whole wax component showed a higher percentage of phenol-like compounds in the resistant genotype than in the susceptibles. Alkylresorcinol content was dramatically higher in GT-MAS:gk wax than in susceptible lines. An alkylresorcinol, 5-methylresorcinol, also inhibited *in vitro* growth of *A. flavus*. Further research is needed for a clear identification of the component(s) responsible for kernel wax resistance and to determine its expression level in other maize lines.

Two Levels of Resistance

The KSA employs a very simple and inexpensive apparatus involving bioassay trays, petri dishes, vial caps as seed containers, and chromatography paper for holding moisture [14]. Kernels screened by the KSA are maintained in 100% humidity, at a temperature favoring *A. flavus* (31° C) growth and aflatoxin production, and are usually incubated for seven days. Aflatoxin data from KSA experiments can be obtained two to three weeks after experiments are initiated. KSA experiments confirmed GT-MAS:gk resistance to aflatoxin production and demonstrated that it is maintained even when the pericarp barrier, in otherwise viable kernels, is breached [13]. Penetration through the pericarp barrier was achieved by wounding the kernel with a hypodermic needle down to the endosperm, prior to inoculation. Wounding facilitates differentiation between different resistance mechanisms in operation, and the manipulation of aflatoxin levels in kernels for comparison with other traits (e.g. fungal growth; protein induction). The results of this study indicate the presence of two levels of resistance: at the pericarp and at the subpericarp level. The former was supported by the above-studies which demonstrated a role for pericarp waxes in kernel resistance [30], and highlighted quantitative and qualitative differences in pericarp wax between GT-MAS:gk and susceptible genotypes [31, 32].

Comparing Fungal Growth to Toxin Production

When selected resistant Illinois maize inbreds (MI82, CI2, and T115) were examined by the KSA, modified to include an *A. flavus* GUS transformant (a strain genetically engineered with a gene construct consisting of a β-glucuronidase reporter gene linked to an *A. flavus* beta-tubulin gene promoter for monitoring fungal growth) [14], kernel resistance to fungal infection in nonwounded and wounded kernels was demonstrated both visually and quantitatively, as was a positive relationship between the degree of fungal infection and aflatoxin levels [14, 33]. This made it possible assesses fungal infection levels and to determine if a correlation exists between infection and aflatoxin levels in the same kernels. *A. flavus* GUS transformants with the reporter gene linked to an aflatoxin biosynthetic pathway gene could also provide a way to indirectly measure aflatoxin levels [34-36], based on the extent of the expression of the pathway gene.

Recently, It was demonstrated, using the KSA and an *F. moniliforme* strain, genetically transformed with a GUS reporter gene linked to an *A. flavus* β-tubulin gene promoter, that the aflatoxin-resistant genotype, GT-MAS:gk, inhibits growth of *F. moniliforme* as well [37]. This indicates that some resistance mechanisms may be generic for ear rotting/mycotoxigenic fungi.

A more recent use of reporter genes was performed on cotton using a green fluorescent protein reporter; a GFP-expressing *A. flavus* strain to successfully monitor fungal growth, mode of entry, colonization of cottonseeds, and production of aflatoxins [38]. This strain provides for an easy, potentially non-destructive, rapid and economical assay which can be done in real time, and may constitute an advance over GUS transformants.

Resistance-Associated Proteins

Developing resistance to fungal infection in wounded as well as intact kernels would go a long way toward solving the aflatoxin problem [17]. Studies demonstrating subpericarp (wounded-kernel) resistance in maize kernels have led to research for identification of subpericarp resistance mechanisms. Examinations of kernel proteins of several genotypes revealed differences between genotypes resistant and susceptible to aflatoxin contamination [39]. Imbibed susceptible kernels, for example, showed decreased aflatoxin levels and contained germination-induced ribosome inactivating protein (RIP) and zeamatin [40]. Both zeamatin and RIP have been shown to inhibit *A. flavus* growth *in vitro* [40]. In another study, two kernel proteins were identified from a resistant corn inbred (Tex6) which may contribute to resistance to aflatoxin contamination [41]. One protein, 28 kDa in size, inhibited *A. flavus* growth, while a second, over 100 kDa in size, primarily inhibited toxin formation. When a commercial corn hybrid was inoculated with aflatoxin and nonaflatoxin-producing strains of *A. flavus*at milk stage, one induced chitinase and one ß-1,3-glucanase isoform was detected in maturing infected kernels, while another isoform was detected in maturing uninfected kernels [42].

In another investigation, an examination of kernel protein profiles of 13 maize genotypes revealed that a 14 kDa trypsin inhibitor protein (TI) is present at relatively high concentrations in seven resistant maize lines, but at low concentrations or is absent in six susceptible lines [43]. The mode of action of TI against fungal growth may be partially due to its inhibition of fungal -amylase, limiting *A. flavus*access to simple sugars [44] required not only for fungal growth, but also for toxin production [45]. TI also demonstrated antifungal activity against other mycotoxigenic species [46]. The identification of these proteins may provide markers for plant breeders, and may facilitate the cloning and introduction of antifungal genes through genetic engineering into other aflatoxin-susceptible crops.

An investigation into maize kernel resistance [47] determined that both constitutive and induced proteins are required for resistance to aflatoxin production. It also showed that one major difference between resistant and susceptible genotypes is that resistant lines constitutively express higher levels of antifungal proteins compared to susceptible lines. The real function of these high levels of constitutive antifungal proteins may be to delay fungal invasion, and consequent aflatoxin formation, until other antifungal proteins can be synthesized to form an active defense system.

Proteomic Analysis

Two-dimensional (2-D) gel electrophoresis, which sorts proteins according to two independent properties, isoelectric points and then molecular weights, has been recognized for a number of years as a powerful biochemical separation technique. Improvements in map resolution and reproducibility [48,49], rapid analysis of proteins, analytical software and computers, and the acquisition of genomic data for a number of organisms has given rise to another application of 2-D electrophoresis: proteome analysis. Proteome analysis or "proteomics" is the analysis of the protein complement of a genome [50,51]. This involves the systematic separation, identification, and quantification of many proteins simultaneously. 2-D electrophoresis is also unique in its ability to detect post- and cotranslational modifications, which cannot be predicted from the genome sequence.

Through proteome analysis and the subtractive approach, it may be possible to identify important protein markers associated with resistance, as well as genes encoding these proteins. This could facilitate marker-assisted breeding and/or genetic engineering efforts. Endosperm and embryo proteins from several resistant and susceptible genotypes have been compared using large format 2-D gel electrophoresis, and over a dozen such protein spots, either unique or 5-fold upregulated in resistant maize lines (Mp420 and Mp313E), have been identified, isolated from preparative 2-D gels and analyzed using ESI-MS/MS after in-gel digestion with trypsin [52, 53]. These proteins, all constitutively expressed, can be grouped into three categories based on their peptide sequence homology: (1) storage proteins, such as globulins and late embryogenesis abundant proteins; (2) stress-responsive proteins, such as aldose reductase, a glyoxalase I protein and a 16.9 kDa heat shock protein, and (3) antifungal proteins, including the above-described TI.

During the screening of progeny developed through the IITA-USDA/ARS collaborative project, near-isogenic lines from the same backcross differing significantly in aflatoxin accumulation were identified, and proteome analysis of these lines is being conducted [54]. Investigating corn lines from the same cross with contrasting reaction to *A. flavus* should enhance the identification of RAPs clearly without the confounding effect of differences in the genetic backgrounds of the lines.

Heretofore, most RAPs identified have had antifungal activities. However, increased temperatures and drought, which often occur together, are major factors associated with aflatoxin contamination of maize kernels [55]. It has also been found that drought stress imposed during grain filling reduces dry matter accumulation in kernels [55]. This often leads to cracks in the seed and provides an easy entry site to fungi and insects. Possession of unique or of higher levels of hydrophilic storage or stress-related proteins, such as the aforementioned, may put resistant lines in an advantageous position over susceptible genotypes in the ability to synthesize proteins and defend against pathogens under stress conditions. Further studies including physiological and biochemical characterization, genetic mapping, plant transformation using RAP genes, and marker-assisted breeding should clarify the roles of stress-related RAPs in kernel resistance. RNAi gene silencing experiments involving RAPs may also contribute valuable information. [54].

Further Characterization of Raps

A literature review of the RAPs identified above indicates that storage and stress-related proteins may play important roles in enhancing stress tolerance of host plants. The expression of storage protein GLB1 and LEA3 has been reported to be stress-responsive and ABA-dependant [56]. Transgenic rice overexpressing a barley LEA3 protein HVA1 showed significantly increased tolerance to water deficit and salinity [57]. The role of GLX I in stress-tolerance was first highlighted in an earlier study using transgenic tobacco plants overexpressing a *Brassica juncea* glyoxalase I [58]. The substrate for glyoxalase I, methylglyoxal, is a potent cytotoxic compound produced spontaneously in all organisms under physiological conditions from glycolysis and photosynthesis intermediates, glyceraldehydes-3-phosphate and dihydroxyacetone phosphate. Methylglyoxal is an aflatoxin inducer even at low concentrations; experimental evidence indicates that induction is through upregulation of aflatoxin biosynthetic pathway transcripts including the *AFLR* regulatory gene [59]. Therefore, glyoxalase I may be directly affecting resistance by removing its aflatoxin-inducing substrate, methylglyoxal. PER1, a 1-cys peroxiredoxin antioxidant identified in a proteomics investigation [60], was demonstrated to be an abundant peroxidase, and may play a role in the removal of reactive oxygen species. The PER1 protein overexpressed in *Escherichia coli* demonstrated peroxidase activity *in vitro*. It is possibly involved in removing reactive oxygen species produced when maize is under stress conditions [60]. Another RAP that has been characterized further is the pathogenesis-related protein 10 (PR10). It showed high homology to PR10 from rice (85.6% identical) and sorghum (81.4% identical). It also shares 51.9% identity to intracellular pathogenesis-related proteins from lily (AAF21625) and asparagus (CAA10720), and low homology to a RNase from ginseng [61]. The PR10 overexpressed in *E. coli* exhibited ribonucleolytic and antifungal activities. In addition, an increase in the antifungal activity against *A. flavus* growth was observed in the leaf extracts of transgenic tobacco plants expressing maize *PR10* gene compared to the control leaf extract [61]. This evidence suggests that PR10 plays a role in kernel resistance by inhibiting fungal growth of *A. flavus*. Further, its expression during kernel development was induced in the resistant line GT-MAS:gk, but not in susceptible Mo17 in response to fungal inoculation [61]. Recently, a new *PR10* homologue was identified from maize (*PR10.1*) [62]. *PR10* was expressed at higher levels in all tissues compared to*PR10.1*, however, purified PR10.1 overexpressed in *E. coli* possessed 8-fold higher specific RNase activity than *PR10* [62]. This homologue may also play a role in resistance. Evidence supporting a role for *PR10* in host resistance is also accumulating in other plants. A barley *PR10* gene was found to be specifically induced in resistant cultivars upon infection by *Rhynchosporium secalis*, but not in near-isogenic susceptible plants [63]. In cowpea, a *PR10* homolog was specifically up-regulated in resistant epidermal cells inoculated with the rust fungus *Uromyces vignae* Barclay [64]. A *PR10* transcript was also induced in rice during infection by *Magnaporthe grisea* [65].

To directly demonstrate whether selected RAPs play a key role in host resistance against *A. flavus*infection, an RNA interference (RNAi) vector to silence the expression of endogenous RAP genes (such as *PR10*, *GLX I* and *TI*) in maize through genetic engineering was constructed [59, 66]. The degree of silencing using RNAi constructs is greater than that obtained using either co-suppression or antisense constructs, especially when an intron is included [67]. Interference of double-stranded RNA with expression of specific genes has been widely described [68, 69]. Although the mechanism is still not well understood, RNAi provides an extremely powerful tool to study functions of unknown genes in many organisms. This posttranscriptional gene silencing (PTGS) is a sequence-specific RNA degradation process triggered by a dsRNA, which propagates systemically throughout the plant, leading to the degradation of homologous RNA encoded by endogenous genes, and transgenes. Both particle bombardment and *Agrobacterium*-mediated transformation methods were used to introduce the RNAi vectors into immature maize embryos. The former was used to provide a quick assessment of the efficacy of the RNAi vector in gene silencing. The latter, which can produce transgenic materials with fewer copies of foreign genes and is easier to regenerate, was chosen for generating transgenic kernels for evaluation of changes in aflatoxin-resistance. It was demonstrated using callus clones from particle bombardment that *PR10* expression was reduced by an average of over 90% after the introduction of the RNAi vector [66]. The transgenic kernels also showed a significant increase in susceptibility to *A. flavus* infection and aflatoxin production. The data from this RNAi study clearly demonstrated a direct role for PR10 in maize host resistance to *A. flavus* infection and aflatoxin contamination [66]. RNAi vectors to silence other RAP genes, such as *GLX I* and *TI*, have also been constructed, and introduced into immature maize embryos through both bombardment and *Agrobacterium* infection [70]. It will be very interesting to see the effect of silencing the expression of these genes in the transgenic kernels on host resistance to *A. flavus* infection and aflatoxin production.

ZmCORp, a protein with a sequence similar to cold-regulated protein and identified in the above-proteomic studies, was shown to exhibit lectin-like hemagglutination activity against fungal conidia and sheep erythrocytes [71]. When tested against *A. flavus*, ZmCORp inhibited germination of conidia by 80% and decreased mycelial growth by 50%, when germinated conidia were incubated with the protein. Quantitative real-time RT-PCR revealed *ZmCORp* to be expressed 50% more in kernels of a resistant maize line *versus* a susceptible.

ZmTIp, a 10 kDa trypsin inhibitor, had an impact on *A. flavus* growth, but not as great as the previously-mentioned 14 kDa TI [72].

Proteomic Studies of Rachis and Silk Tissue

A study was conducted to investigate the proteome of rachis tissue, maternal tissue that supplies nutrients to the kernels [75]. An interesting finding in this study is that after infection by *A. flavus*, rachis tissue of aflatoxin-resistant genotypes did not up-regulate PR proteins as these were already high in controls where they had strongly and constitutively accumulated during maturation. However, rachis tissue of aflatoxin-susceptible lines did not accumulate PR proteins to such an extent during maturation, but increased them in response to fungal infection. Given the relationship of the rachis to kernels, these results confirm findings of a previous investigation [47], which demonstrated levels of proteins in resistant *versus* susceptible kernels was a primary factor that determined kernel genetic resistance to aflatoxin contamination. Another study was conducted to identify proteins in maize silks that may be contributing to resistance against *A. flavus* infection/colonization

[76]. Antifungal bioassays were performed using silk extracts from two aflatoxin-resistant and two–susceptible inbred lines. Silk extracts from resistant inbreds showed greater anti-fungal activity compared to susceptible inbreds. Comparative proteomic analysis of the two resistant and susceptible inbreds led to the identification of antifungal proteins including three chitinases that were differentially-expressed in resistant lines. When tested for chitinase activity, silk proteins from extracts of resistant lines also showed significantly higher chitinase activity than that from susceptible lines. Differential expression of chitinases in maize resistant and susceptible inbred silks suggests that these proteins may contribute to resistance.

Transcriptomic Analyses

To investigate gene expression in response to *A. flavus'* infection and to more thoroughly identify factors potentially involved in the regulation of RAP genes, a transcriptomic profile was conducted on maize kernels of two inbred lines that were genetically closely-related [73]. Similar work had previously been performed using Tex6 as the resistant line and B73 as the susceptible [74], however, in the study using closely-related lines, imbibed mature kernels were used (for the first time) and proved to be a quicker and easier approach than traditional approaches. The involvement of certain stress-related and antifungal genes previously shown to be associated with constitutive resistance was demonstrated here; a kinase-binding protein, Xa21 was highly up-regulated in the resistant line compared to the susceptible, both constitutively and in the inducible state.

CURRENT EFFORTS TO DEVELOP RESISTANT LINES

Closely-Related Lines

Recently, the screening of progeny generated through a collaborative breeding program between IITA-Nigeria (International Institute of Tropical Agriculture) and the Southern Regional Research Center of

USDA-ARS in Center (SRRC) of USDA-ARS in New Orleans facilitated the identification of closely-related lines from the same backcross differing significantly in aflatoxin accumulation, and proteome analysis of these lines is being conducted [77, 78]. Investigating corn lines sharing close genetic backgrounds should enhance the identification of RAPs without the confounding effects experienced with lines of diverse genetic backgrounds. The IITA-SRRC collaboration has attempted to combine resistance traits of U.S. resistant inbred lines with those of African lines, originally selected for resistance to ear rot diseases and for potential aflatoxin-resistance (*via* KSA) [77, 78]. Five elite tropical inbred lines from IITA adapted to the Savanna and mid-altitude ecological zones of West and Central Africa were crossed with four U.S. resistant maize lines in Ibadan, Nigeria. The five African lines were originally selected for their resistance to ear rot caused by *Aspergillus*, *Botrydiplodia*,*Diplodia*, *Fusarium*, and *Macropomina* [77, 78]. The F1 crosses were backcrossed to their respective U.S. inbred lines and self-pollinated thereafter. The resulting lines were selected through the S4 generation for resistance to foliar diseases and desirable agronomic characteristics under conditions of severe natural infection in their respective areas of adaptation. Promising S5 lines were screened with the KSA (Table 1). In total, five pairs of closely-related lines were shown to be significantly different in aflatoxin resistance, while sharing as high as 97% genetic similarity [79]. Using these lines in proteomic comparisons to identify RAPs has advantages: (1) gel comparisons and analyses become easier; and (2) protein differences between resistant and susceptible lines as low as twofold can be identified with confidence. In addition, the likelihood of identifying proteins that are directly involved in host resistance is increased. In a preliminary proteomics comparison of constitutive protein differences between those African closely-related lines, a new category of resistance-associated proteins (putative regulatory proteins) was identified, including a serine/threonine protein kinase and a translation initiation factor 5A [29, 79]. The genes encoding these two resistance associated regulatory proteins are being cloned and their potential role in host resistance to *A. flavus* infection and aflatoxin production will be further investigated. Conducting proteomic analyses using lines from this program not only enhances chances of identifying genes important to resistance, but may have immediate practical value. The IITA-SRRC collaboration has registered and released six inbred lines with aflatoxin-resistance in good agronomic backgrounds, which also demonstrate good levels of resistance to southern corn blight and southern corn rust [80]. Resistance field trials for these lines on U.S. soil is being conducted; the ability to use resistance in these lines commercially will depend on having identified excellent markers, since seed companies desire insurance against the transfer of undesirable traits into their elite genetic backgrounds. The fact that this resistance is coming from good genetic backgrounds is also a safeguard against the transfer of undesirable traits.

Table 1: KSA screening of IITA-SRRC maize breeding materials which identified 2 closely related lines (87.5% genetic similarity), #22 and #25, from parental cross (GT-MASgk x Ku1414SR) x GT-MAS:gk; these contrast significantly in aflatoxin accumulation. Values followed by the same letter are not significantly different by the least significant difference test (P = 0.05)

Entry	Aflatoxin B_1 (ppb)
Susceptible control	10197 a
22*	1693 b
19	1284 bc
28	1605 bcd
27	1025 bcd
21	1072 bcd
26	793 bcde
20	574 cde
24	399 cde
GT-MAS:gk	338 de
25*	228 e
23	197 e
Resistant control	76 e

Recent Breeding Efforts

Recent breeding efforts towards the development of aflatoxin-resistant maize lines has resulted in a number of germplasm releases including the above-mentioned IITA-SRRC inbreds. In 2008, TZAR 101-106, derived from a combination of African and southern-adapted U.S. lines are being field-tested in different parts of the Southern U.S. (Figure

1) [80]. These have also exhibited resistance to lodging and common foliar diseases. GT-603 was released in 2011, after having been derived from GT-MAS:gk [81], while Mp-718 and Mp-719 were released as southern adapted resistant lines which are both shorter and earlier than previous Mp lines [82, 83]. These lines are also being tested as inbreds and in hybrid combinations in the southern U.S. [83].

Figure 1: Inoculation of maize ears with *Aspergillus flavus* spores using a 'side needle' wound technique for field evaluations of TZAR lines developed through IITA-SRRC program.

CONCLUSION

The host resistance approach to eliminating aflatoxin contamination of maize has been advanced forward by the identification/development of maize lines with resistance to aflatoxin accumulation. However, to fully exploit the resistance discovered in these lines, markers must be identified to transfer resistance to commercially useful backgrounds. Towards this goal numerous investigations have been undertaken to discover the factors that contribute to resistance, laying the basis for exploiting these discoveries as well. These investigations include QTL analyses to locate regions of chromosomes associated with the resistant phenotype, and the discovery of kernel resistance-related traits. We now know that there are two levels of resistance in kernels, pericarp and subpericarp. Also, there is a two-phased kernel resistance response to fungal attack: constitutive at the time of fungal attack and that which is induced by the attack. Thus far, it's been demonstrated that natural resistance mechanisms discovered are antifungal in nature as opposed to inhibiting the aflatoxin biosynthetic pathway.

One of the most important discoveries, thus far, has been that of resistance-associated proteins or RAPs. Due to the significance of the constitutive response, constitutive RAPs were investigated first, although induced proteins are being studied as well. Investigations of other tissues such as rachis and silks begin to provide a more complete picture of the maize resistance response to aflatoxigenic fungi. RAP characterization studies provide greater evidence that these proteins are important to resistance, although clearly, more investigations are needed. Looking at data collectively that's been obtained from different types of studies may enhance the identification of markers for breeding. A good example of this may be the supporting evidence provided by QTL data to proteomic and RAP characterization data suggesting the involvement of 14 kDa TI, water stress inducible protein, zeamatin, heat shock, cold-regulated, glyoxalase I, cupin-domain and PR10 proteins in aflatoxin-resistance. It will be interesting to determine if this marker discovery approach can lead to the successful transfer of a multigene-based and quantitative phenomenon such as aflatoxin-resistance to commercially-useful genetic backgrounds.

ACKNOWLEDGEMENTS

Research discussed in this review received support from the USAID Linkage Program-IITA, Nigeria, and the USDA-ARS Office of International Research Programs (OIRP) -USAID Collaborative Support Program.

REFERENCES

1. Brown RL., Bhatnagar D., Cleveland TE., Cary JW. Recent Advances in Preventing Mycotoxin Contamination. In: Sinha KK., Bhatnagar D. (eds) Mycotoxins in Agriculture and Food Safety, Marcel Dekker: New York, NY, USA; 1998. p351-379.
2. Dorner JW., Cole RJ., Wicklow DT. Aflatoxin Reduction in Maize through Field Application of Competitive Fungi. Journal of Food Protection 1999; 62 650–656.
3. CAST. Aflatoxins and Other Mycotoxins: An Agricultural Perspective CAST Report No. 80; Council for Agricultural Science and Technology: Ames, IA; 1979.
4. Diener UL., Cole RJ., Sanders TH., Payne GA., Lee LS. Epidemiology of Aflatoxin Formation by *Aspergillus flavus*. Annual Review of Phytopathology 1987; 25 249–270.
5. Haumann F. Eradicating Mycotoxins in Food and Feeds. Inform 1995; 6 248–256.
6. Gong YY., Cardwell K., Hounsa A., Egal S., Turner PC., Hall AJ.,Wild CP. Dietary Aflatoxin Exposure and Impaired Growth in Young Children from Benin and Togo: cross sectional study. British Medical Journal 2002; 325 20–21.
7. Probst C., Njapau H., Cotty PJ. Outbreak of an Acute Aflatoxicosis in Kenya in 2004: Identification of the Causal Agent. Applied and Environmental Microbiology 2007; 73 2762–2764.
8. Gardner CAC., Darrah LL., Zuber MS., Wallin JR. Genetic Control of Aflatoxin Production in Maize. Plant Disease 1987;71 426–429.
9. King SB., Scott GE. Screening Maize Single Crosses for Resistance to Preharvest Infection of Kernels by *Aspergillus flavus*. Phytopathology 1982;72 942.
10. Widstrom NW., McMillian WW., Wilson DM. Segregation for Resistance to Aflatoxin Contamination Among Seeds on an Ear of Hybrid Maize. Crop Science 1987;27 961–963.
11. Scott GE., Zummo N. Sources of Resistance in Maize to Kernel Infection by *Aspergillus flavus* in the Field. Crop Science 1988;28 505–507.

12. Campbell KW., White DG. (1995). Evaluation of Maize Genotypes for Resistance to Aspergillus Ear Rot, Kernel Infection, and Aflatoxin Production. Plant Disease 1995;79 1039–1045.
13. Brown RL., Cotty PJ., Cleveland TE., Widstrom NW. (1993). Living Maize Embryo Influences Accumulation of Aflatoxin in Maize Kernels. Journal of Food Protection 1993; 56 967–971.
14. Brown RL., Cleveland TE., Payne GA., Woloshuk CP., Campbell KW., White DG. Determination of Resistance to Aflatoxin Production in Maize Kernels and Detection of Fungal Colonization Using an *Aspergillus flavus* Transformant Expressing *Escherichia coli* β-glucuronidase. Phytopathology 1995; 85 983–989.
15. King SB., Wallin JR. Methods for screening corn for resistance to kernel infection and aflatoxin production by Aspergillus flavus. In:Diener, UL., Asquith RL., Dickens JW. (eds) Aflatoxin and *Aspergillus flavus* in Corn (Southern Cooperative Bulletin 279). Auburn University, Alabama, 1983; p77.
16. Tucker DH Jr., Trevathan LE., King SB., Scott GE. Effect of Four Inoculation Techniques on Infection and Aflatoxin Concentration of Resistant and Susceptible Corn Hybrids Inoculated with *Aspergillus flavus*. Phytopathology 1986; 76 290-293.
17. Payne GA. Aflatoxin in Maize. Critical Reviews in Plant Science 1992; 10 423-440.
18. Shotwell O. Aflatoxin detection and determination in corn. In: Diener UL.,Asquith RL., Dickens JW. (eds) Aflatoxin and *Aspergillus flavus* in Corn (Southern Cooperative Bulletin 279)., Auburn University, Alabama, 1983; p38.
19. Windham GL, Williams WP. *Aspergillus flavus* Infection and Aflatoxin Accumulation in Resistant and Susceptible Maize Hybrids. Plant Disease 1998; 82 281-284.
20. White DG., Rocheford TR., Kaufman B., Hamblin AM. Chromosome regions associated with resistance to *Aspergillus flavus* and inhibition of aflatoxin production in maize. In: Proceedings of USDA-ARS Aflatoxin Elimination Workshop, Atlanta, GA, USA, 1995; p.8.
21. White DG., Rocheford TR., Naidoo G., Paul C., Hamblin AM. Inheritance of molecular markers associated with, and breeding for resistance to Aspergillus Ear Rot and aflatoxin production

in corn using Tex6. In: Proceedings of USDA-ARS Aflatoxin Elimination Workshop, St. Louis, MO, USA, 1998; pp. 4–6.

22. Paul C., Naidoo G., Forbes A., Mikkilineni V., White D., Rocheford T. Quantitative Trait Loci for Low Aflatoxin Production in Two Related Maize Populations. Theoretical and Applied Genetics 2003; 7 263–270.

23. Busboom KN. White DG. Inheritance of Resistance to Aflatoxin Production and Aspergillus Ear Rot of Corn from the Cross of Inbreds B73 and Oh516. Phytopathology 2004; 94 1107–1115.

24. Brooks TD., Williams WP., Windham GL., Wilcox MC., Abbas H. Quantitative Trait Loci Contributing Resistance to Aflatoxin Accumulation in Maize Inbred Mp313E. Crop Science 2005; 45 171–174.

25. Davis GL., Windham GL., Williams WP. QTL for Aflatoxin Reduction in Maize. Maize Genetics Conference Abstracts 2000; 41 139.

26. Warburton ML., Brooks TD., Krakowsky MD., Shan X., Windham GL., Williams WP. Identification and Mapping of New Sources of Resistance to Aflatoxin Accumulation in Maize. Crop Science 2009; 49 1403–1408.

27. Warburton ML., Brooks TD., Windham GL., Williams WP. Identification of Novel QTL Contributing Resistance to Aflatoxin Accumulation in Maize. Molecular Breeding 2011; 27 491-499.

28. Brown RL., Chen Z-Y., Luo M., Menkir A., Fakhoury A., Bhatnagar D. Discovery and Characterization of Proteins Associated with Aflatoxin-Resistance: Building a Case for Their Potential as Breeding Markers. Toxins 2010 2(4) 919-933.

29. Chen Z-Y., Brown R L., Menkir A., Cleveland TE. Identification of Resistance-Associated Proteins in Closely-Rrelated Maize Lines Varying in Aflatoxin Accumulation. Molecular Breeding 2012; 30 53-68.

30. Guo BZ., Russin JS., Cleveland TE., Brown RL., Widstrom NW. Wax and Cutin Layers in Maize Kernels Associated with Resistance to Aflatoxin Production by *Aspergillus flavus*. Journal of Food Protection 1995; 58 296-300.

31. Gembeh SV., Brown RL., Grimm C., Cleveland TE. Identification of Chemical Components of Corn Kernel Pericarp Wax Associated

with Resistance to *Aspergillus flavus* Infection and Aflatoxin Production. Journal of Agricultural and Food Chemistry 2001; 49 4635-4641.

32. Russin JS., Guo BZ., Tubajika KM., Brown RL., Cleveland TE., Widstrom NW. Comparison of Kernel Wax from Corn Genotypes Resistant or Susceptible to *Aspergillus flavus*. Phytopathology 1997; 87 529-533.
33. Brown RL., Cleveland TE., Payne GA., Woloshuk CP., White DG. Growth of an *Aspergillus flavus* Transformant Expressing *Escherichia coli* ß- glucuronidase in Maize Kernels Resistant to Aflatoxin Production. Journal of Food Protection 1997; 60 84-87.
34. Brown RL., Brown-Jenco CS., Bhatnagar D., Payne GA. Construction and preliminary evaluation of an *Aspergillus flavus* reporter gene construct as a potential tool for screening aflatoxin resistance. Journal of Food Protection 2003a; 66 1927-1931.
35. Brown-Jenco C S., Brown R L., Bhatnagar D., Payne G A. 1998, Use of an omtA(p)::GUS reporter construct to evaluate corn for resistance to aflatoxin accumulation. Proceedings of the USDA-ARS Aflatoxin Elimination Workshop, St. Louis, 1998; pp. 23.
36. Payne GA. 1997. Characterization of inhibitors from corn seeds and the use of a new reporter construct to select corn genotypes resistant to aflatoxin accumulation. Proceedings of the USDA-ARS Aflatoxin Elimination Workshop, Memphis, TN, 1997; pp 66-67.
37. Brown RL., Cleveland T E., Woloshuk CP., Payne GA., Bhatnagar D. Growth Inhibition of a *Fusarium verticillioides* GUS Strain in Corn Kernels of Aflatoxin -Resistant Genotypes. Applied Microbiology and Biotechnology 2001a; 57 708-711.
38. Rajasekaran K., Cary JW., Cotty PJ., Cleveland TE. Development of a GFP-Expressing *Aspergillus flavus* Strain to Study Fungal Invasion, Colonization, and Resistance in Cottonseed. Mycopathologia 2008; 165 89-97.
39. Guo BZ., Brown RL., Lax AR., Cleveland TE., Russin JS., Widstrom NW. Protein Profiles and Antifungal Activities of Kernel Extracts from Maize Genotypes Resistant and Susceptible to *Aspergillus flavus*. Journal of Food Protection 1998; 61 98–102.

40. Guo BZ., Chen Z-Y., Brown RL., Lax AR., Cleveland TE., Russin JS., Mehta A.D., Selitrennikoff CP., Widstrom NW. Germination Induces Accumulation of Specific Proteins and Antifungal Activities in Maize Kernels. Phytopathology 1997; 87 1174–1178.

41. Huang Z., White DG., Payne GA. Maize Seed Proteins Inhibitory to *Aspergillus flavus* and Aflatoxin Biosynthesis. Phytopathology 1997; 87 622–627.

42. Ji C., Norton RA., Wicklow DT., Dowd PE. Isoform Patterns of Chitinase and β-1,3-glucanase in Maturing Maize Kernels (*Zea mays* L.) Associated with *Aspergillus flavus* Milk Stage Infection. Journal of Agricultural and Food Chemistry 2000; 48 507–511.

43. Chen Z-Y., Brown RL., Lax AR., Guo BZ., Cleveland TE., Russin JS. Resistance to *Aspergillus flavus* in Maize Kernels is Associated with a 14 kDa Protein. Phytopathology 1998; 88 276–281.

44. Chen Z-Y., Brown RL., Russin JS., Lax AR., Cleveland TE (1999b). A Maize Trypsin Inhibitor with Antifungal Activity Inhibits *Aspergillus flavus* α-amylase. Phytopathology 1996b; 89 902–907.

45. Woloshuk CP., Cavaletto JR., Cleveland TE. Inducers of Aflatoxin Biosynthesis from Colonized Maize Kernels are Generated by an Amylase Activity from *Aspergillus flavus*. Phytopathology 1997; 87 164–169.

46. Chen Z-Y., Brown RL., Lax AR., Cleveland TE., Russin JS. Inhibition Plant Pathogenic Fungi by a Maize Trypsin Inhibitor Over-Expressed in*Escherichia coli*. Applied and Environmental Microbiology 1999a; 65 1320–1324.

47. Chen ZY., Brown RL., Cleveland TE., Damann KE., Russin JS. Comparison of Constitutive and Inducible Maize Kernel Proteins of Genotypes Resistant or Susceptible to Aflatoxin Production. Journal of Food Protection 2001; 64 1785–1792.

48. GorgA., PostelW., GuntherS., WeserJ., Improved Horizontal Two-Dimensional Electrophoresis with Hybrid Isoelectric Focusing in Immobilized pH Gradients in the First Dimension and Laying-On Transfer to the Second Dimension. Electrophoresis 1985; 6 599-604.

49. Gorg A., Postel W., Gunther S.,The Current State of Two-Dimensional Electrophoresis with Immobilized pH Gradients. Electrophoresis 1988; 9 531-546.

50. Pennington SR., Wilkins MR., Hochstrasser DF., Dunn MJ. Proteome Analysis: From Protein Characterization to Biological Function. Trends in Cell Biology 1997; 7 168-173.

51. Wilkins MR., Pasquali C., Appel RD., Ou K., Golaz O., Sanchez JC., Yan JX., Gooley AA., Hughes G., Humphrey-Smith I., Williams KL., Hochstrasser DF. From Proteins to Proteomes: Large-Scale Protein Identification by Two-Dimensional Electrophoresis and Amino Acid Analysis. Biotechnology 1996; 14 61-65.

52. Chen Z-Y., Brown RL., Damann KE., Cleveland TE. 2000. Proteomics analysis of kernel embryo and endosperm proteins of corn genotypes resistant or susceptible to *Aspergillus flavus* infection. In: Proceedings of the USDA-ARS Aflatoxin Elimination Workshop, 2000; pp. 88.

53. Chen Z-Y., Brown R.L., Damann K.E., Cleveland T.E. Identification of Unique or Elevated Levels of Kernel Proteins in Aflatoxin-Resistant Maize Genotypes through Proteome Analysis. Phytopathology 2002; 92 1084–1094.

54. Brown RL., Chen Z-Y., Menkir A., Cleveland TE. Using biotechnology to enhance host resistance to aflatoxin contamination of corn. African Journal of Biotechnology 2003b; 2 557-562.

55. Payne GA. Process of Contamination by Aflatoxin-Producing Fungi and Their Impact on Crops. In: Sinha KK., Bhatnagar D. (eds), Mycotoxins in Agriculture and Food Safety. Marcel Dekker, New York, NY, USA, 1998; pp. 279-306.

56. Thomann E.B., Sollinger J., White C., Rivin C.J. Accumulation of Group 3 Late Embryogenesis Abundant Proteins in *Zea mays* Embryos. Plant Physiology 1992; 99 607–614.

57. Xu D., Duan X., Wang B., Hong B., Ho THD., Wu R. Expression of a Late Embryogenesis Abundant Protein Gene HVA1, from Barley Confers Tolerance to Water Deficit and Salt Stress in Transgenic Rice. Plant Physiology 1996; 110 249–257.

58. Veena V., Reddy S., Sopory SK. Glyoxalase I from Brassica juncea: Molecular Cloning, Regulation and Its Over-expression Confer

Tolerance in Transgenic Tobacco under Stress. Plant Journal 1999; 17 385–395.

59. Chen ZY., Brown RL., Damann KE., Cleveland TE. Identification of a Maize Kernel Stress-Related Protein and Its Effect on Aflatoxin Accumulation. Phytopathology 2004; 94 938–945.

60. Chen ZY., Brown RL., Damann KE., Cleveland TE. Identification of Maize Kernel Endosperm Proteins Associated with Resistance to Aflatoxin Contamination by *Aspergillus flavus*. Phytopathology 2007; 97 1094–1103.

61. Chen Z.Y., Brown RL., Rajasekaran K., Damann KE., Cleveland TE. Evidence for Involvement of a Pathogenesis-Related Protein in Maize Resistance to *Aspergillus flavus* Infection/Aflatoxin Production. Phytopathology 2006; 96 87–95.

62. Xie YR., Chen Z-Y., Brown RL., Bhatnagar D. Expression and Functional Characterization of Two Pathogenesis-Related Protein10 Genes from *Zea mays*. Journal of Plant Physiology 2010; 167 121–130.

63. Steiner-Lange S., Fischer A., Boettcher A., Rouhara I., Liedgens H., Schmelzer E., Knogge W. Differential Defense Reactions in Leaf Tissues of Barley in Response to infection by *Rhynchosporium secalis* and to Treatment with a Fungal Avirulence Gene Product. Molecular and Plant-Microbe Interactions 2003; 16 893–902.

64. Mould M.J., Xu T., Barbara M., Iscove NN., Heath MC. cDNAs Generated from Individual Eepidermal Cells Reveal that Differential Gene Expression Predicting Subsequent Resistance or Susceptibility to Rust Fungal Infection Occurs Prior to the Fungus Entering the Cell Lumen. Molecular and Plant-Microbe Interactions 2003; 16 835–845.

65. McGee JD., Hamer JE., Hodges TK. Characterization of a PR-10 Pathogenesis-Related Gene Family Induced in Rice during Infection with*Magnaporthe grisea*. Molecular and Plant-Microbe Interactions 2001; 14 877–886.

66. Chen Z-Y., Brown RL., Damann KE., ClevelandTE. PR10 Expression in Maize and Its Effect on Host Resistance against *Aspergillus flavus*/infection/Aflatoxin Production. Molecular Plant Pathology 2010; 11 69–81.

67. Wesley SV., Helliwell CA., Smith NA., Wang MB., Rouse DT., Liu Q., Gooding PS., Singh SP., Abbott D., Stoutjesdijk PA. Robinson SP., Gleave AP., Green AG., Waterhouse PM. Construct design for efficient, effective and high-throughput gene silencing in plants. Plant Journal 2001; 27 581–590.

68. Fire A. Xu S., Montgomery MK., Kostas SA., Driver SE., Mello CC. Potent and Specific Genetic Interference by Double Stranded RNA in*Caenorhabditis elegans*. Nature 1998; 391 806–811.

69. Gura T. A silence that speaks volumes. Nature 2000; 404 804–808.

70. Chen Z-Y., Damann KE. Louisiana State University, Baton Rouge, LA; Brown RL., Cleveland TE. USDA-ARS-SRRC New Orleans, LA., Unpublished work. 2007.

71. Baker R., Brown RL., Cleveland TE., Chen Z-Y., Fakhoury A. COR, a Maize Lectin-like Protein with Antifungal Activity against *Aspergillus flavus*. Journal of Food Protection 2009a; 72 120–127.

72. Baker R., Brown RL., Cleveland TE., Chen Z-Y., Fakhoury A. ZmTI, a Maize Trypsin Inhibitor with Limited Activity against *Aspergillus flavus*. Journal of Food Protection 2009b; 72 85–188.

73. Luo M., Brown R L., Chen Z-Y., Menkir A., Yu J., Bhatnagar D. Transcriptional Profiles Uncover *Aspergillus flavus* Induced Resistance in Maize Kernels. Toxins 2011; 3 766-785.

74. Luo M., Liu J., Lee RD., Scully BT., Guo B. Monitoring the Expression of Maize Genes in Developing Kernels under Drought Stress Using Oligo-Microarray. Journal of Integrative Plant Biology 2010; 52(12) 1059-1074.

75. Pechanova O., Pechan T., Williams, WP., Luthe DS. Proteomic Analysis of Maize Rachis: Potential Roles of Constitutive and Induced Proteins in Resistance to Aspergillus flavus Infection and Aflatoxin Contamination. Proteomics 2011; 11 114-127.

76. Peethambaran B., Hawkins L., Windham GL., Williams WP., Luthe DS. Antifungal Activity of Maize Silk Proteins and Role of Chitinases in*Aspergillus flavus* Resistance. Toxin Reviews 2010; 29 27–39.

77. Brown RL., Chen Z-Y., Menkir A., Cleveland T.E., Cardwell K., Kling, J., White D.G. Resistance to Aflatoxin Accumulation in

Kernels of Maize Inbreds Selected for Ear Rot Resistance in West and Central Africa. Journal of Food Protection 2001b; 64 396–400.

78. Menkir A., Brown RL., Bandyopadhyay R.., Chen Z-Y., Cleveland TE. A USA-Africa Collaborative Strategy for Identifying, Characterizing, and Developing Maize Germplasm with Resistance to Aflatoxin Contamination. Mycopathologia 2006; 162 225–232.

79. Chen ZY., Brown RL., Menkir A., Damann KE., Cleveland TE. Proteome Analysis of Near Isogenic Maize Lines Differing in the Level of Resistance against *Aspergillus flavus* Infection/Aflatoxin Production. Phytopathology 2005; 95 S19.

80. Menkir A., Brown RL., Bandyopadhyay R., Cleveland TE. Registration of Six Tropical Maize Germplasm Lines with Resistance to Aflatoxin Contamination. Journal of Plant Registrations 2008; 2 246–250.

81. Guo BZ., Krakowsky MD., Ni X., Scully BT., Lee RD., Coy AE., Widstrom NW. Registration of Maize Inbred Line GT603. Journal of Plant Registrations 2011; 5(2) 211-214.

82. Williams WP.,Windham GL. Registration of Mp718 and Mp719 Germplasm Lines of Maize. Journal of Plant Registrations 2012; 6 200-202.

83. Scully BT., Guo BZ., Ni X., Williams WP., Henry WB., Krakowsky MD., Brown RL. Development of Aflatoxin and Insect Resistant Corn Inbreds Adapted to the Southern U.S. Proceedings of the Corn Utilization and Technology Workshop, Abstract (in press).

Chapter 5

Heterosis in Early Maize Ear Inflorescence Development: A Genome-Wide Transcription Analysis for two Maize Inbred Lines and Their Hybrid

Haiping Ding,[1] Cheng Qin,[1,2]; Xirong Luo,[2] Lujiang Li,[1] Zhe Chen,[1] Hongjun Liu,[1] Jian Gao,[1] Haijian Lin,[1] Yaou Shen,[1] Maojun Zhao,[3] Thomas Lübberstedt,[4] Zhiming Zhang,[1] and Guangtang Pan[1]

[1]Maize Research Institute of Sichuan Agricultural University/Key Laboratory of Biology and Genetic Improvement of Maize in Southwest Region, Ministry of Agriculture, Chengdu 611130, China;

[2]Zunyi Academy of Agricultural Sciences, Zunyi 563102, China;

[3]Life Science College, Sichuan Agricultural University, Ya'an 625014, China;

[4]Department of Agronomy, Iowa State University, Ames, IA 50011, USA;

ABSTRACT

Heterosis, or hybrid vigor, contributes to superior agronomic performance of hybrids compared to their inbred parents. Despite its importance, little is known about the genetic and molecular basis of heterosis. Early maize ear inflorescences formation affects grain yield, and are thus an excellent model for molecular mechanisms involved in heterosis. To determine the parental contributions and their regulation during maize ear-development-genesis, we analyzed genome-wide digital gene expression profiles in two maize elite inbred lines (B73 and Mo17) and their F_1 hybrid using deep sequencing technology. Our analysis revealed 17,128 genes expressed in these three genotypes and 22,789 genes expressed collectively in the present study. Approximately 38% of the genes were differentially expressed in early maize ear inflorescences from heterotic cross, including many transcription factor genes and some presence/absence variations (PAVs) genes, and exhibited multiple modes of gene action. These different genes showing differential expression patterns were mainly enriched in five cellular component categories (organelle, cell, cell part, organelle part and macromolecular complex), five molecular function categories (structural molecule activity, binding, transporter activity, nucleic acid binding transcription factor activity and catalytic activity), and eight biological process categories (cellular process, metabolic process, biological regulation, regulation of biological process, establishment of localization, cellular component organization or biogenesis, response to stimulus and localization). Additionally, a significant number of genes were expressed in only one inbred line or absent in both inbred lines. Comparison of the differences of modes of gene action between previous studies and the present study revealed only a small number of different genes had the same modes of gene action in both maize seedlings and ear inflorescences. This might be an indication that in different tissues or developmental stages, different global expression patterns prevail, which might nevertheless be related to heterosis. Our results support the hypotheses that multiple molecular mechanisms (dominance and overdominance modes) contribute to heterosis.

INTRODUCTION

Heterosis, or hybrid vigor, refers to the phenomenon in which progeny of two inbred lines (hybrids) exhibit enhanced agronomic performance such as biomass production, growth rate, fertility, and disease resistance relative to both parents [1]. Heterosis has been extensively used in agriculture, especially in the breeding of maize and rice. For example, it is estimated that approximately 95% of the United States maize acreage and 55% of rice acreage in China are planted with hybrids. Furthermore, hybrid maize technology for large-scale production has a yield advantage of 15% over the elite inbred varieties. Though the concept of heterosis has been introduced over 100 years ago and different genetic models considered [1,2,3], the genetic and molecular basis of heterosis remains elusive.

Three classical genetic hypotheses to explain heterosis have been proposed: the dominance, overdominance, and epistasis hypotheses. The dominance hypothesis states that deleterious recessive alleles cause inbreeding depression. A cross of two inbred parents will benefit from complementation of these deleterious alleles and will display a superior phenotype [2,3,4]. The over-dominance hypothesis refers to allelic interactions at one or multiple heterozygous genes resulting in superior trait expression compared to the better parent [1,5]. Both hypotheses may be insufficient to explain the molecular mechanism for heterosis [6]. Epistasis hypothesis refers to interactions of alleles at different loci from two parents in F_1 hybrids, leading to heterosis [7,8].

Recently, genetic and molecular studies provided experimental support for above-mentioned hypotheses. Using molecular markers in segregating populations, quantitative trait locus (QTL) mapping studies provided support for the dominance [9], over-dominance [10], and epistasis [7,8] models. With the advent of genomic methods to assay genome-wide patterns of gene expression, recent studies indifferent tissues and developmental stages of model and crop plants (such as maize, rice, wheat, *Populus*, and *Arabidopsis*) have determined the roles of different gene expression [11,12,13,14,15,16,17,18,19,20,21,22,23,24,25,26,27,28,29,30,31,32], small RNAs [33,34,35,36], and epigenetic regulation [37,38], including circadian-mediated metabolic pathways [39], in heterosis. For instance, additive gene expression was prevalent in some studies of maize [14,26,32], and there were varying numbers of genes that exhibited a non-additive behavior in other studies [12,17,18,19,20,21,22,23,24,25,26,27,29,30,35].

Maize immature ear inflorescences show heterosis in ear architectural traits [40], and significant positive correlations between grain yield and ear architectural traits, such as ear length, kernel row number, number of kernels per row, kernel number density, and cob diameter have been reported [40], which suggests that each of these components contributes to greater yields. A thorough knowledge of the genes affecting the various components and their interactions will facilitate our understanding of the molecular basis of heterosis of grain yield. In this study, we applied a highly effective approach of high throughout deep sequencing to identify genes, which are highly expressed in maize elite inbred lines of B73 and Mo17, and their F_1 hybrid (B73 × Mo17) at an early stage of ear inflorescences development. We provide first molecular evidence that regulatory mechanisms underlying the phenomenon of heterosis are very early active in maize ear development. Furthermore, we also found that differentially expressed genes between hybrids and their parents can be involved in certain regulatory networks, which suggested that complicated gene networks might be underlying heterosis. Results of the present study might help promote further understanding of mechanisms underlying heterosis.

RESULTS

Statistics and Analysis of Library Sequencing

Here, we sequenced three ear digital gene expression (DGE) libraries from two inbred parents (B73 and Mo17) and their F_1 hybrid (B73 × Mo17) using massively parallel sequencing on the Illumina platform at BGI-Shenzhen, China (Table 1 and Table S1). A total of approximately 4.2 million raw tags per library with 259,890 distinct tag sequences were identified. After data-processing steps (see Materials and Methods), the total number of filtered, high-quality clean tags was almost the same in three libraries. Furthermore, the F_1 library had the highest number of distinct tags (259,282), followed by the B73 (242,184), and Mo17 (239,963) libraries (Table 1 and Table S1). These distinct tags and their genomic frequency as well as the raw data were deposited in NCBI Sequence Read Archive (SRA) database with the accession number (PRJNA248701). Copy numbers of most of the distinct tags (over 77%) ranged from 1–5. However, a small number of distinct tags (less than 3.29%) with a frequency higher than 100 make up over 62% of all clean tags in all three libraries (Figure S1).

Table 1: Summary statistics from mapping digital gene expression (DGE) sequencetags to the maize B73 reference genome

Class	Summary	F_1	B73	Mo17
Raw Data	Total	4,200,000	4,200,000	4,200,000
	Distinct Tags	272,402	254,882	252,386
Clean Tags	Total number	4,176,622	4,176,825	4,176,752
	Distinct Tags number	259,282	242,184	239,963
All Tags Mapping to Genome	Total number	2,766,685	2,979,615	3,127,520
	Total % of clean tags	66.24%	71.34%	74.88%
	Distinct Tags number	134,449	134,378	141,476
	Distinct Tags % of clean tags	51.85%	55.49%	58.96%
Unambiguous Tags Mapping to Genome	Total number	2,452,293	2,637,419	2,804,389
	Total % of clean tags	58.71%	63.14%	67.14%
	Distinct Tags number	119,316	118,719	125,467
	Distinct Tags % of clean tags	46.02%	49.02%	52.29%

All Tags Mapping to Genes	Number	24,629	24,078	24,198
	% of ref. genes	75.69%	74%	74.36%
Unambiguous Tags Mapping to Genes	Number	21,372	20,784	20,938
	% of ref. genes	65.68%	63.87%	64.35%
Unknown Tags	Total number	894,503	721,786	561,476
	Total % of clean tags	21.42%	17.28%	13.44%
	Distinct Tags number	85,152	70,700	61,016
	Distinct Tags % of clean tags	32.84%	29.19%	25.43%

When sequencing depths reach 1 million total tags, the number of novel distinct tags discovered dropped dramatically in all three libraries (Figure S2A). From that point, increasing sequence depth results in a slow and stable accumulation of new distinct tags indicating that sequencing has reached saturation. Moreover, as shown in Figure 2B, when the total tag number in B73 reached 1 million, the increase of identified genes started to level out, and stabilized when the number of tags reached 3 million. Next, the level of gene expression was determined by calculating the number of unambiguous clean tags for each gene and then normalized to the number of transcripts per million tags (TPM). The Mo17 and F_1 data showed a similar trend (Figure S2). This suggests that only few more distinct genes would be identified when the total clean tag number reached a certain value.

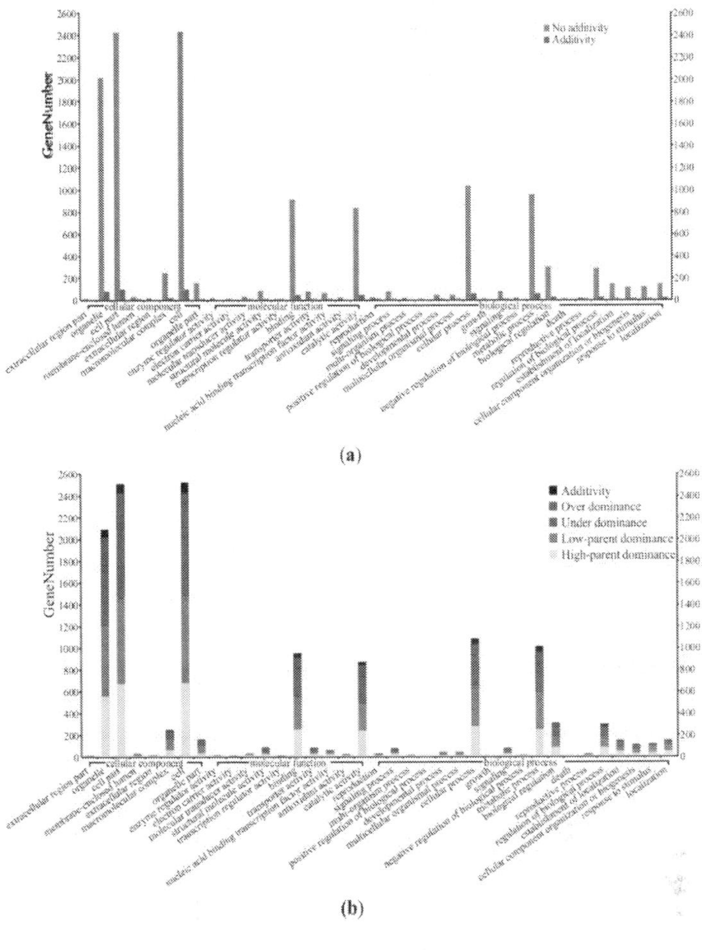

Figure 1: Functional categories of differentially expressed genes. Overall differentially expressed genes (a) and non-additive expressed genes (b) from B73 × Mo17 cross.

Mapping Tags to the Maize Reference Genome

We used SOAP2 software [41] to map all distinct tags to the maize reference genome (B73 RefGen_v2) [42]. Mapping results showed that 51.85%, 55.49% and 58.96%, respectively, of distinct clean tags mapped to the reference database (sense or anti-sense), and 46.02%,

49.02%, and 52.29%, respectively, of the distinct clean tags mapped unambiguously to the reference genes (Table 1 and Table S1). Out these, we identified 20,784–21,372 genes expressed in the three genotype comparison (Table S2), our analysis revealed 17,128 genes expressed in all samples and 22,789 genes expressed collectively in the present study (Table S3). In total, 32.84%, 29.19%, and 25.43% of all distinct clean tags for F_1, B73, and Mo17 data sets, respectively, did not map to the reference maize genome sequence or associated transcripts (Table 1 and Table S1). These non-mapped tags most likely represent regions where the maize reference sequence is incomplete [43] or there are differential mRNA processing events for most maize genes, such as alternative splicing [44]. Only 0.03% of non-mapped tags matched maize chloroplast or mitochondrial genome sequences (Table S1). Because Solexa sequencing can distinguish transcripts originating from both DNA strands, we found evidence for bidirectional transcription in 14,012–14,420 of all detectable overlapping genes and 1031–1119 antisense-stand specific transcripts based on the strand-specific nature of the sequenced tags (Table S2). By comparison, the ratio of sense to antisense strand of the transcripts was approximately 1.3:1 for all libraries. As summarized in Figure S3 and Table S3, most expressed genes (approx. 19,965) are represented in fewer than one hundred copies and only a small proportion of genes are highly expressed (Table S3). We map the clean tags, which cannot be mapped to mRNA, mitochondria and chloroplasts, to the whole genome, providing start positions that could be uniquely mapped by those tags (Table S4).

Different Gene Expression Analysis

A total of 32,973 significantly changed tags entities were detected among three genotypes (see Materials and Methods) (Figure S4A,C,E and Table S5). Then, the processed DGE data were used to determine different gene expression between inbred parents or between parental lines and their F_1 hybrid and we identified 8621 out of 22,789 genes that were differentially expressed among B73, Mo17, and their hybrid, representing 38% of the ear digital gene expression (Figure 1A). Among them, 3401 (3344) and 2226 (2189) genes were significantly up-regulated and down-regulated, respectively, in the F_1 compared to B73/Mo17 (Figure 1B,Figure S4B,D and Table S6). The comparison between parental line libraries also revealed significant variation in expression.

A total of 3883 genes, including 1896 up-regulated and 1987 down-regulated genes, were identified in B73 compared to Mo17 (Figure 1B, Figure S4F and Table S6).

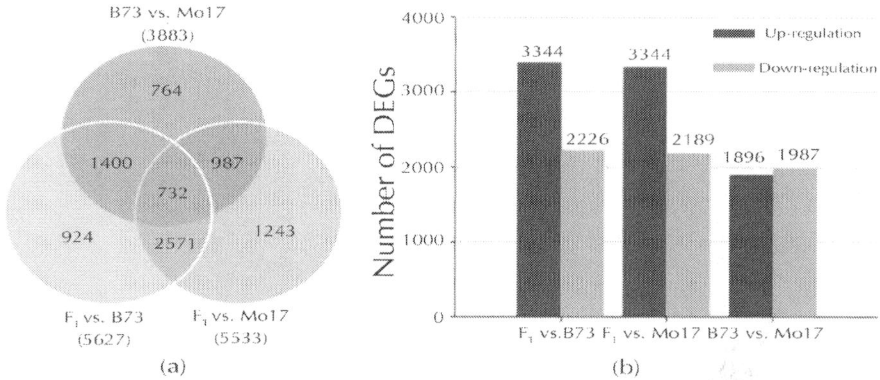

Figure 2: Differentially expressed genes in the maize heterotic cross. Venn diagram (a) and statistical analysis (b) of differentially expressed genes among inbred parents and their F_1 hybrid.

We further investigated the mode of gene action for these different genes (Table S6). 8.9% (767 of 8621) exhibited an expression pattern that was not distinguishable from additivity, while the other 91.1% (7854 of 8621) genes showed non-additive expression patterns (Figure 2A). The non-additive differentially expressed genes from the cross were further classified into four distinct classes: high-parent dominance (1984), low-parent dominance (2559), over-dominance (1085), and under-dominance (1963) (Figure 2B and Table 2). A sample of 30 differentially expressed genes was randomly selected for validation by qRT-PCR. The trends in the expression of these genes detected by DGE were consistent (29 genes) or partially (1 gene) consistent with those determined in qRT-PCR analyses (Figure 3). These findings are consistent with a recent study in maize [31] or rice [16] that supports the involvement of multiple modes (dominance and over dominance) of gene action in association with heterosis.

Table 2: Statistical analysis of differentially expressed genes (FDR ≤ 0.001 and absolute value of $\log_2$Ratio ≥ 1)

Hybrid Cross	**B73 × Mo17**
Total	8621
Additivity	767
Non Additivity	7854
High-Parent Dominance	1984
Low-Parent Dominance	2559
Over Dominance	1085
Under Dominance	1963
Other	263

F_1 represents the hybrid line; P, paternal line, B73; and M, maternal line, Mo17. Additivity, $F_1 \approx 1/2(P + M)$; non-additivity, $F_1 > 1/2\ (P + M)$ or $F_1 < 1/2\ (P + M)$. High-parent dominance (HPD), $F_1 \approx P > M$ or $F_1 \approx M < P$; low-parent dominance (LPD), $F_1 \approx P < M$ or $F_1 \approx M < P$; over-dominance (ODO), $F_1 > P$ and $F_1 > M$; under-dominance (UDO), $F_1 < P$ and $F_1 < M$

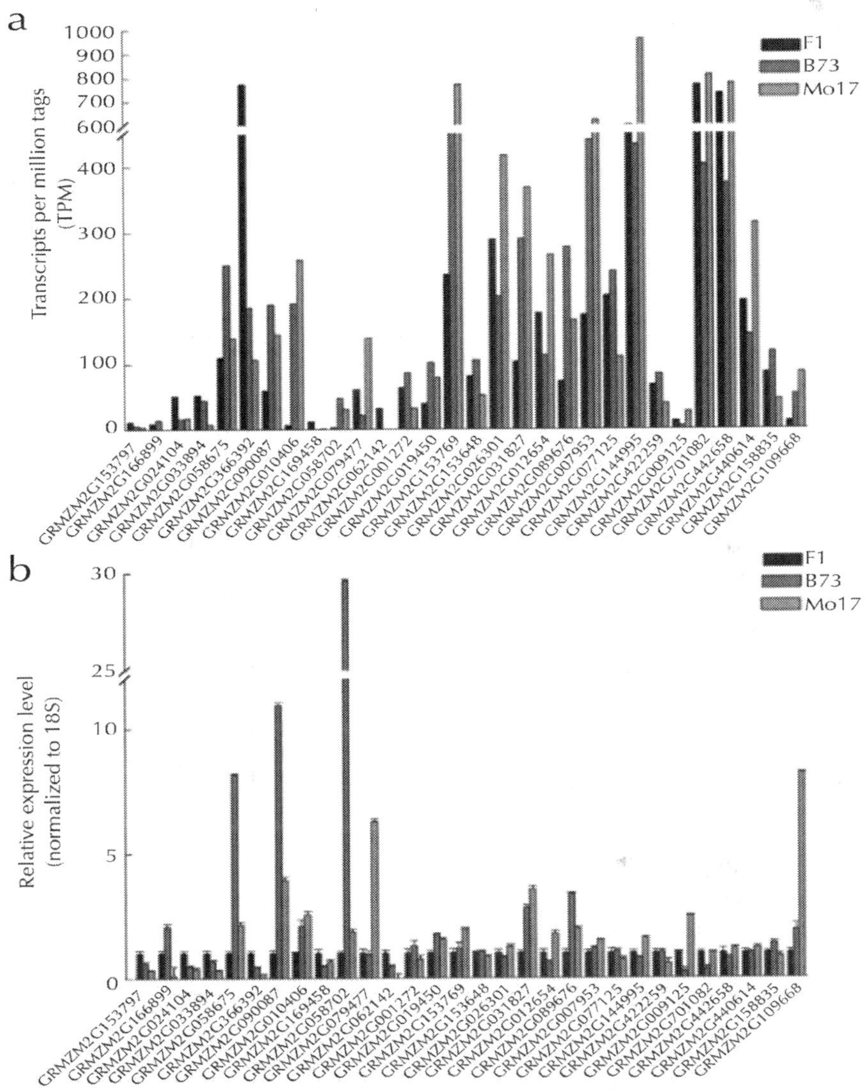

Figure 3: Expression of 12 differentially expressed gene from arginine and proline metabolism, pyruvate metabolism, and purine metabolism in the maize heterotic cross. The expression pattern of 12 genes detected by digital gene expression platform(a) and quantitative Real-Time PCR (qRT-PCR) (b). The black, red, and green bars in graph b depicted the stem-loop qRT-PCR relative expression level ± standard error of three replicates for each gene inB73, Mo17 and their hybrid.

Functional Enrichment Analysis for Differentially Expressed Genes Using Gene Ontology (GO)

To gain a better understanding of the functional roles of different genes between inbred parents or between parental lines and their F_1 hybrid, we looked for gene enrichment regarding the Gene Ontology (GO) cellular component, molecular function and biological process categories [44,45]. Functional-annotations from the maize genome (http://maizesequence.org) were used for functional classification of the 8621 different genes and performed by the official web-based tools for searching and browsing the Gene Ontology database (AmiGO) (http://www.geneontology.org/) [45] (Figure 2 and Table 3). We found that different genes showing differential expression patterns were mainly enriched in five cellular component categories (organelle, cell, cell part, organelle part and macromolecular complex), five molecular function categories (structural molecule activity, binding, transporter activity, nucleic acid binding transcription factor activity and catalytic activity), and eight biological process categories (cellular process, metabolic process, biological regulation, regulation of biological process, establishment of localization, cellular component organization or biogenesis, response to stimulus and localization) (Table 3). We further classified different genes in more detail based on Gene Ontology (GO) by AmiGO [45] (Table S7–S9). These different genes in GO annotation analysis of cellular component showed significant enrichment in the following components: non-membrane-bound organelle (6.40×10^{-4}), intracellular non-membrane-bound organelle (6.40×10^{-4}), nucleus (2.56×10^{-3}), ribosome (3.34×10^{-2}), and macromolecular complex (4.27×10^{-2}). In the GO biological process enrichment analysis, there are five significant GO terms (FDR corrected p-value < 0.05). Eighty percent were macromolecular synthesis-related, such as cellular protein metabolic process, cellular macromolecule metabolic process, protein metabolic process, and macromolecule metabolic process (Table 4).

Table 3: Functional classification of different genes between inbred parents or between parental lines and their F_1 hybrid.* indicated the significant enrichment in functional classification

Functional Categories	Additivity	Non-Additivity	HPD	LPD	UDO	ODO	Other
Cellular Component							
Extracellular Region Part	1	3	0	2	0	1	0
Organelle *	181	2077	544	650	542	271	70
Cell Part	221	2499	655	793	624	339	88
Membrane-Enclosed Lumen	1	24	3	9	10	1	1
Extracellular Region	1	13	1	5	2	5	0
Macromolecular Complex *	16	245	58	89	71	16	11
Cell	221	2510	660	794	628	340	88
Organelle Part	16	155	31	57	47	14	6
Molecular Function							
Enzyme Regulator Activity	3	12	3	2	3	4	0
Electron Carrier Activity	0	3	0	1	2	0	0
Molecular Transducer Activity	3	24	7	5	9	2	1
Structural Molecule Activity	3	80	17	27	28	4	4
Transcription Regulator Activity	2	6	0	1	2	2	1
Binding	75	940	241	297	258	108	36
Transporter Activity	9	76	25	19	20	9	3

Nucleic Acid Binding Transcription Factor Activity	2	58	18	19	15	2	4
Antioxidant Activity	0	17	5	3	3	6	0
Catalytic Activity	67	865	229	263	226	111	36
Biological Process							
Reproduction	0	21	3	8	5	3	2
Signaling Process	5	70	24	20	16	7	3
Multi-Organism Process	0	8	4	2	1	1	0
Positive Regulation of Biological Process	0	3	2	1	0	0	0
Developmental Process	2	38	2	21	9	4	2
Multicellular Organismal Process	2	39	4	20	9	4	2
Cellular Process	71	1079	269	357	281	125	47
Growth	0	5	1	2	1	1	0
Signaling	5	70	24	20	16	7	3
Negative Regulation of Biological Process	2	12	2	6	3	1	0
Metabolic Process	70	999	242	338	253	118	48
Biological Regulation	18	303	77	94	88	28	16
Death	2	6	2	1	2	1	0
Reproductive Process	0	15	1	5	5	2	2
Regulation of Biological Process	17	289	76	88	84	26	15
Establishment of Localization	15	143	41	44	33	19	6

Cellular Component Organization or Biogenesis	10	106	24	38	29	10	5
Response to Stimulus	4	112	33	37	22	19	1
Localization	16	145	41	45	33	19	7

Table 4: Significant GO terms of DGs in the gene ontology (GO) annotation analysis of cellular component and biological process. [a]*p*-values calculated using a hypergeometric test-determines if the number of times that a GO term appears in the cluster is significant, relative to its occurrence in the genome

GO Term	Definition	FDR Corrected *p*-Value [a]
Cellular Component		
GO:0043228	Non-membrane-bounded organelle	6.40×10^{-4}
GO:0043232	Intracellular non-membrane-bounded organelle	6.40×10^{-4}
GO:0005634	Nucleus	2.56×10^{-3}
GO:0005840	Ribosome	3.34×10^{-2}
GO:0032991	Macromolecular complex	4.27×10^{-2}
Biological Process		
GO:0044267	Cellular protein metabolic process	4.22×10^{-3}
GO:0044260	Cellular macromolecule metabolic process	8.84×10^{-3}
GO:0019538	Protein metabolic process	3.30×10^{-2}
GO:0009987	Cellular process	3.37×10^{-2}
GO:0043170	Macromolecule metabolic process	4.59×10^{-2}

The Expression Patterns of Biological Macromolecular Synthesis-Related Genes in the B73 × Mo17 Hybrid

To investigate pathways in which different genes were involved and enriched, pathway analysis was performed using the KOBAS 2.0 web tool [46]. 2935 out of 5627 different genes were involved in 122 pathways between F_1 and B73, 2877 out of 5533 different genes were involved in 121 pathways between $F_1$and Mo17 (Table S10), Among them, the majority were present in ribosome, spliceosome and various metabolic pathways (such as arginine and proline metabolism, purine metabolism, glycolysis/gluconeogenesis, and biosynthesis of alkaloids derived from histidine and purine, *etc.*) (Table S10). *q*-value estimates [47] revealed only four pathways showing significance ($q < 0.05$): ribosome ($q = 1.58 \times 10^{-3}$), arginine and proline metabolism ($q = 1.58 \times 10^{-3}$), spliceosome ($q = 1.63 \times 10^{-2}$) and pyruvate metabolism ($q = 3.80 \times 10^{-2}$) (Table 5 and Table S11), and the first two pathways showed extreme significance ($q < 0.01$). Of the top 10 differentially expressed genes enriched in pathways between parental lines and their F_1 hybrid, it should be noted that the purine metabolism pathway was the third largest pathway, following the ribosome pathway, although it does not reach the significance level (Table 5 and Table S11).

Table 5: Top 10 pathways differentially expressed genes enriched in between parental lines and their F_1 hybrid

Pathway [a]	Different Genes with Pathway Annotation (4312)				*p* -Value [b]	*q* -Value [c]	Pathway ID
	F_1 vs. B73	**F_1 vs. Mo17**	**Overlap**	**Total (%)**			
Ribosome **	148	129	89	188 (4.68%)	2.47×10^{-5}	1.58×10^{-3}	ko03010
Arginine and Proline Metabolism **	37	34	25	46 (1.11%)	2.59×10^{-5}	1.58×10^{-3}	ko00330
Spliceosome *	145	136	87	194 (4.82%)	4.01×10^{-4}	1.63×10^{-2}	ko03040
Pyruvate Metabolism *	32	41	25	49 (1.69%)	6.28×10^{-4}	3.80×10^{-2}	ko00620
Proteasome	33	25	19	39 (0.90%)	2.00×10^{-3}	6.09×10^{-2}	ko03050
Pentose Phosphate Pathway	16	24	13	27 (0.63%)	1.66×10^{-3}	6.70×10^{-2}	ko00030
Purine Metabolism	64	72	39	97 (2.25%)	9.34×10^{-4}	7.50×10^{-2}	ko00230
RNA Degradation	43	49	31	61 (1.41%)	4.38×10^{-3}	8.84×10^{-2}	ko03018
Porphyrin and Chlorophyll Metabolism	21	16	11	26 (0.60%)	3.82×10^{-3}	9.32×10^{-2}	ko00860

Glycolysis/ Gluconeogenesis	49	57	31	75 (1.74%)	8.38 × 10^{-3}	1.04 × 10^{-1}	ko00010

[a]Pathwayanalysisbased on KOBAS 2.0 (Mao *et al.* 2005; Wu *et al.* 2006; Xie *et al.* 2011) [46,48,49];[b]*p*-value in hypergeometric test; [c]The *q*-value is similar to the well-known *p*-value, except it is a measure of significance in terms of the false discovery rate rather than the false positive rate (Storey *et al.* 2003) [47]; the top 10 pathways with *q*-value are listed; * pathway with *q*-value < 0.05 is considered as significant; ** pathway with *q*-value < 0.01 is considered as extreme significant.

In the four significant pathways, 188 different genes (4.68%) were detected in the ribosome pathway, 46 different genes (1.11%) in the arginine and proline metabolism pathway, 194 (4.82%) in the spliceosome pathway and 49 (1.69%) in the pyruvate metabolism pathway. Approximately 50% of different genes in every pathway were differentially expressed between inbred and hybrid (Table 5). In the two significant metabolic pathways, almost all (202 out of 204) different genes showed non-additive expression patterns (HPD, UDO, LPD, or UDO). Seventeen of the 46 different genes in the arginine and proline metabolism pathway were up-regulated (with expression pattern of HPD or ODO) in the F_1 hybrid, and 26 of 46 were down-regulated (with expression pattern of LPD or UDO). As compared with parental lines, the transcriptional level of one different gene in the F_1 hybrid (GRMZM2G062142) encoding ornithine carbamoyltransferase (OTC) was up-regulated by 7- and 11.6-fold, respectively. The gene (GRMZM2G169458) encoding fatty aldehyde dehydrogenase 1 and another gene (GRMZM2G366392) encoding *S*-adenosylmethionine decarboxylase proenzyme (SAMDC) showed 4.7-/3.4-fold and 2.1-/2.9-fold higher expression in F_1 hybrid than the parental lines, respectively. Simultaneously, the expression levels of another different gene in the F_1 hybrid (GRMZM2G035042) encoding IMP dehydrogenase/GMP reductase were down-regulated by 8.5- and 9.3-fold, respectively, comparing with parental lines, another gene (GRMZM2G010406) encoding argininosuccinate synthase showed significant down-regulation expression in F_1 hybrid. Moreover, 50 different genes involved in the pyruvate metabolism included genes encoding glyoxylatereductase (GRMZM2G166899), phosphoenolpyruvate carboxylase1 (PEPCase 1, GRMZM2G083841), oxidoreductase (GRMZM2G118770), and Pyruvate, orthophosphate dikinase 1 (GRMZM2G097457). (Table S8).

Resolving Transcription Factors (TFs) among Differentially Expressed Genes

A primary objective was to identify genes that encode TFs and to determine their modes of gene action. To test this, we retrieved putative orthologs of maize genes in our differently expressed data based on information from the Ensembl Compara gene trees [50] at maizesequence.org and gramene.org [51]. We then queried known *Arabidopsis* TFs in the database of *Arabidopsis* Transcription Factors (http://datf.cbi.pku.edu.cn/) and identified 435 maize different genes with sequence similarities to *Arabidopsis*TFs between parental lines and their F_1 hybrid libraries (Table S12). We further interrogated these different genes using gene ontologies, Inter Pro domains, and known maize annotations. Of the 435 putative TFs, 11 exhibited additive gene action, and the majority of these TFs (n = 424) detected in this study exhibited non additive modes of gene expression. Most of these genes exhibited low-parent dominance (n = 138), high-parent dominance (n = 112), and under dominance (n = 112). However, over dominance (n = 42) and other gene action (n = 19) were also observed.

We also surveyed the differential expression of TFs across a wide range of transcript abundance in hybrid here and a mutant in *RAMOSA3* (*RA3*) gene reported in a previous study [52], the latter of which showed an increased branching phenotype resulting from a loss of determinacy of basal spikelet pair meristems. A total of 39 differentially expressed putative TFs in our dataset were identified, and exhibited all non-additive modes of gene expression (Table 6). Moreover, these TFs were also differentially expressed over a wide range of abundances in *ra3* mutants [52]. These TFs includes several kinds of members of TF families associated with functions in development and meristem maintenance or identity (NAC, YABBY, GRAS and TCP), while others have roles in hormone-mediated or stress-mediated signaling by auxin (AUX/IAA), brassinosteroids (BES), or ethylene and stress (AP2/ERF). Among the 42 differentially expressed TFs, eight were characterized as AP2/ERF family proteins (Table 6). Therefore, these

TFs possibly not only contribute to heterosis, but also provide insight into genetic control of branching.

Table 6: Differentially expressed maize genes were identified as putative TFs

Maize Gene ID	Annotation [a]	TF Family	Significant Pattern [b]	Expression Model [c]	Regulated in *ra3* Mutants [d]
GRMZM2G106673	B3-domain TF	B3	Mo17 < F_1 ≈ B73	HPD	Up
GRMZM2G177046	Ocs element-binding factor 1	bZIP	Mo17 < F_1 ≈ B73	HPD	Up
GRMZM2G102514	BES1/BZR1 protein	BES	Mo17 < F_1 ≈ B73	HPD	Down
GRMZM2G172657	Uncharacterized	GRAS	Mo17 < F_1 ≈ B73	HPD	Down
GRMZM2G115357	IAA24	AUX/IAA	Mo17 < F_1 ≈ B73	HPD	Up
GRMZM2G181376	Uncharacterized	–	B73 < F_1 ≈ Mo17	HPD	Down
GRMZM2G173534	Inducer of CBF expression 2	–	B73 < F_1 ≈ Mo17	HPD	Up
GRMZM2G173124	Zinc finger	Znf-C3H1	B73 ≈ F_1 < Mo17	LPD	Up
GRMZM2G138886	Cyclin B2	Cyclin	B73 ≈ F_1 < Mo17	LPD	Up
GRMZM2G020054	Uncharacterized	AP2/ERF	Mo17 ≈ F_1 < B73	LPD	Up
GRMZM2G055243	KNOX class 2 protein	KNOX	Mo17 ≈ F_1 < B73	LPD	Up
GRMZM2G089995	Ethylene responsive	AP2/ERF	Mo17 ≈ F_1 < B73	LPD	Up
GRMZM2G079825	Pathogenesis-related	AP2/ERF	Mo17 ≈ F_1 < B73	LPD	NS
GRMZM2G310368	Ethylene responsive	AP2/ERF	Mo17 ≈ F_1 < B73	LPD	Up
GRMZM2G132185	Pathogenesis-related	AP2/ERF	Mo17 ≈ F_1 < B73	LPD	NS
GRMZM2G149940	B3 DNA binding domain	B3	Mo17 ≈ F_1 < B73	LPD	Up
GRMZM2G003927	Ramosa1	Znf-C2H2	B73 ≈ F_1 < Mo17	LPD	Up
GRMZM2G447406	Progesterone receptor	–	B73 ≈ F_1 < Mo17	LPD	Down
GRMZM2G102218	YABBY protein	YABBY	Mo17 ≈ F_1 < B73	LPD	Up

GRMZM2G136769	Ubiquitin-associated	Ubiquitin	Mo17 ≈ F_1 < B73	LPD	Up
GRMZM2G165972	Heat shock factor (HSF)-type	HSF	Mo17 ≈ F_1 < B73	LPD	Down
GRMZM2G140474	Tyrosine protein kinase	–	Mo17 ≈ F_1 < B73	LPD	Up
GRMZM2G422205	Zinc finger	Znf-C3H1	Mo17 ≈ F_1 < B73	LPD	
GRMZM2G171468	Uncharacterized	MYB	Mo17 ≈ F_1 < B73	LPD	Down
AC206951.3_FG017	Ethylene-responsive element binding protein 2	ERF	Mo17 < B73 < F_1	ODO	Up
GRMZM2G081012	Transcription initiation factor IID, 18 kD subunit family protein	TFIID-18	B73 ≈ Mo17 < F_1	ODO	Up
GRMZM2G014653	NAC protein 48	NAC	B73 ≈ Mo17 < F_1	ODO	Up
GRMZM2G127379	NAM containing	NAC	B73 ≈ Mo17 < F_1	ODO	Up
GRMZM2G061487	DRE binding factor 1	AP2/ERF	F_1 < Mo17 < B73	UDO	Up
GRMZM2G431157	Zinc finger, C_2H_2-type	Znf_C2H2-like	F_1 < B73 ≈ Mo17	UDO	Up
GRMZM2G307119	Branched silkless1	AP2/ERF	F_1 < B73 ≈ Mo17	UDO	Up
GRMZM2G144275	bHLH transcription factor	HLH	F_1 < B73 ≈ Mo17	UDO	Up
GRMZM2G132367	HDZipI-1	HD-Zip	F_1 < B73 ≈ Mo17	UDO	Up
GRMZM2G105266	Pathogenesis-related	AP2/ERF	F_1 < B73 ≈ Mo17	UDO	NS
GRMZM2G381395	DNA-directed RNA polymerase	–	F_1 < B73 ≈ Mo17	UDO	Down
GRMZM2G453424	Uncharacterized	HRDC_like	F_1 < B73 ≈ Mo17	UDO	Up
GRMZM2G118113	DNA-directed RNA polymerase II 8.2 kDa polypeptide	EF	F_1 < B73 ≈ Mo17	UDO	Up
GRMZM2G017606	SHI	SHI	F_1 < B73 ≈ Mo17	UDO	Up
GRMZM2G039889	Cold acclimation protein	–	F_1 < B73 ≈ Mo17	UDO	Up
GRMZM2G088309	Drooping leaf	YABBY	F_1 < B73 ≈ Mo17	UDO	Up

GRMZM2G078077	TCP domain protein	TCP	Mo 17 < F_1 < B73	Other	Up
GRMZM2G404426	Zinc finger	Znf-PHD	B73 < F_1 < Mo17	Other	Up

[a] Annotations are based on Ensembl gene descriptions at maizesequence.org, gene build5b.60; [b] FDR $\leq$ 0.001 and the absolute value of $\log_2$Ratio $\geq$ 1; [c] Gene expression of F_1 hybrid is classified into multiple patterns. HPD indicates high-parent dominance; LPD, low-parent dominance; ODO, over-dominance; UDO, under-dominance; ND, undistinguishable model from additivity and non-additivity; [d] reference Eveland *et al.* (2010) [52]. NS, Not significant.

A Significant Number of Genes Were Expressed in Only One Inbred Line or Absent in both Inbred Lines

A total of 5660 (24.8%) genes with no detectable expression in one inbred line or both two inbred lines were identified (Table 7). In other word, there are a significant number of genes that displayed presence–absence expression patterns. Among these genes, 46.4% (2624 of 5660) genes exhibited an expression pattern that was present in B73, and absent in Mo17, while some other 34.8% (1971 out of 5660) genes exhibited a similar expression pattern that present in Mo17, and absent in B73 (Table 7 and Table S13). Moreover, these genes expressed in only one inbred line also displayed presence–absence expression patterns in their hybrid. Specifically, regarding the genes present in B73 and absent in Mo17, the ratio of genes present to genes absent in their hybrid was approximately 0.56:1, while the ratio of genes present to genes absent in their hybrid was approximately 1.31:1 for other genes present in Mo17 and absent in B73. In combination, these results suggested that the majority of genes that were expressed in only one inbred line exhibited parental effects on gene expression levels, and presence–absence expression patterns of some genes may be related with presence/absence variations in maize genes [53] (See the following analysis). Surprisingly, it was found that 18.8% (1065 of 5660) genes were not expressed in the two inbred parents. However, these genes were expressed in their hybrid (Table 7 and Table S13). Additionally, nine genes expressed in only one inbred line were also found in previous study by Stupar *et al.* [14], and these results in two studies were consistent. For instance, GRMZM2G037255 (corresponding accession #CF629797) and GRMZM2G152258 (accession #BM073080) absent in both inbred lines in our study were not detected in both B73 and Mo17 by PCR in the study of Stupar *et al.* [14] (Table S13). However, the other 106 genes were expressed in only one inbred in Stupar's study [14], and were expressed in both B73 and Mo17 in present study.

Table 7: Genes with no detectable expression in one inbred line or both two inbred lines

Class	No. of Genes Present in B73, Absent in Mo17	% of Genes	No. of Genes Present in Mo17, Absent in B73	% of Genes	No. of Genes Absent in both B73 and Mo17	% of Genes
No. of genes present in their hybrid	942	16.6	1117	19.7	1065	18.8
No. of genes absent in their hybrid	1,682	29.7	854	15.1	0	0
Total	2,624	46.4	1971	34.8	1065	18.8

Analysis of Presence/Absence Variation (PAVs) Genes Action in Maize Hybrids

Presence/absence variations (PAVs) have been described in maize genes [53], and most of the PAVs reflected true differences in gene content between the B73 and Mo17 genomes in recent research. Because of the availability of a complete list of PAV genes identified by Lai *et al.* [54], the B73 × Mo17 cross was first examined regarding PAV genes between the two parental lines and its relation to different gene expression among the three genotypes. As shown in Table S14, there were 104 PAV genes between the two parental lines, and only 37 PAV genes (35.6%) were mapped by identified tags in three libraries. Interestingly, most of these mapped genes products were hypothetical proteins, and the other 55.8% (58 of 104) of genes were unknown proteins except for nine genes including terpene synthase and ZCN20 (Table S14). Of these mapped genes, 45.9% (17 of 37) genes were expressed in only B73. Six genes expressed in only one inbred line were also not expressed in their hybrid. What was more puzzling is there were three genes which were not expressed in the two inbred parents, nevertheless, expressed in their hybrid. Interestingly, among the 37 PAV genes, the expression levels (TPM) of 30% genes in F_1 hybrid was the same as that of B73 or Mo17, and 50% showed higher or lower expression in F_1 hybrid than both the parental lines (Table S15). However, only five PAV genes were identified with significantly differential expression with HPD or LPD (Table S14).

DISCUSSION

In this study, we assayed genome-wide patterns of gene expression of the maize ear at an early flower differentiation stage among two maize elite inbred lines (B73 and Mo17) and their F_1 hybrid (B73 × Mo17) using Solexa's digital gene expression (DGE) system, a tag-based novel high-throughput transcriptome deep sequencing method. Given the nature of the DGE system, we have pooled biological replicates from three varieties for each group to make representative samples for deep sequencing analysis. We obtained a sequencing depth of approximately 4.2 million tags per library (Table 1) and found 22,789 genes expressed collectively except for putative new transcripts found in the study. Levels of some genes not expressed in the present study were responsive to abiotic stress, e.g., aquaporin PIP1-6 gene (GRMZM2G136032), MADS-box transcription factor 14 gene (GRMZM2G137510), beta-fructofuranosidase 1 precursor gene (GRMZM2G394450), *etc.*, were induced by heavy metal Pb treatment in one recent study [55]. Interestingly, we found evidence for bidirectional transcription in all datasets. By comparison of all libraries, the ratio of sense to antisense strands of the transcripts was approximately 1.3:1, which suggested that not only a high number of antisense expressions, but also the transcriptional regulation in the young ear development acted most strongly on the sense strand. A similar observation was also reported in a recent study [56]. Furthermore, approximately 20% of genes with no detectable expression in one inbred line were identified. These genes may exhibit parental effects on gene expression levels.

Based on our digital gene expression analysis, approximately 37.8% of genes exhibited differential expression between every two genotypes in the B73 × Mo17 hybrid cross. QPCR validation was performed both on the same pooled material that was used for deep sequencing and on independent RNA extractions from each sample, and almost all confirmed the direction of change detected by DGE analysis (Figure 3). These different genes exhibited additive and non-additive expression patterns (Figure 2). A small fraction of DGs (8.9%, 767 of 8621) exhibited a mode of gene action that was indistinguishable from additivity, which is similar to recent studies in maize [14,23,25,31]. Several studies reported more non additively expressed genes, including many with F_1 expression levels outside the parental range [11,12,13,23,27]. Among those non additively expressed genes, the proportion of genes with clear over and under-dominant gene action were 12.6% and 22.8%, respectively, which is similar to results from prior studies [12,15,16,31,57]. Additionally, we further compared the differences of modes of gene action between Swanson-Wagner's study [31] and the present study. A total of 8621 different gene BLAST searches were performed using 1367 ESTs as queries. Four hundred and twenty-one different genes were matched by 547 ESTs with high-scoring segment pairs (Table S15), but only a small number of these genes (75/421) had the same mode of gene action between the two studies. Thus, different global expression patterns in different tissues or developmental stages might prevail. Our results support that multiple molecular mechanisms (dominance and over dominance modes) contribute to heterosis, which is consistent with previous reports [21,31].

No consensus has yet been reached about the genetic basis of heterosis [58]. However, some mechanisms were supported by the observations that sequence polymorphism in promoter alleles between inbred lines preferentially occurred in those differentially transcribed genes [16]. When two alleles are exposed to a common *trans*-acting factor, *cis*-elements in hybrids might differentially interact with gene regulators, resulting in allele-specific gene expression [14,16,59]. This undoubtedly is one of the causes of gene-expression changes in hybrids. There was evidence that phenotypes in hybrids resulted from the dosage effect of such regulatory genes [60]. In this study, we found 424 putative TFs, exhibiting differential expression in the hybrid compared with either parent, in agreement with recent studies

[16]. Remarkably, about 9.2% of these TFs were also differentially expressed in *ra3* mutants (Table 6) and many of the differentially expressed genes that could be mapped onto metabolic pathways were associated with primary carbohydrate biosynthesis and degradation, respiration, and energy production as well as redox and nitrogen cycling processes [52]. In conclusion, the expression of TFs in maize hybrids might be important for allele-specific gene expression in heterosis. Another important finding in a recent study is that many SNPs, indel polymorphisms (IDPs) and PAV genes identified between the B73 and Mo17 genome [54] were consistent with the occurrence of insertion/deletion (indel) variants in 5› regions between the alleles of genes that are differentially expressed in different rice strains [16]. We also analyzed the expression of all 104 PAV genes between the two parental lines by DGE data. Of these genes, 67 PAV genes (64.4%) did not express in both B73 and Mo17, possibly because these genes were nonfunctional. For another 37 PAV genes, most (33/37) of their gene products were hypothetical proteins, unknown proteins, or no significant BLAST hits were obtained by using an e-value cutoff of 1 × 10^{-5}, and almost 50% of these genes were expressed in both inbred lines and their hybrid. Specially, eight PAV genes were expressed only in one inbred line and their hybrid, which possibly contributed to heterosis, because the phenomenon conformed to the previous assumption that inbred lines with large differences in gene content could complement one another [54]. Moreover, the expression levels of approximately 30% (9 of 37) of these genes in the F_1 hybrid was the same as that of B73 or Mo17, and 50% showed higher or lower expression in F_1 hybrid than both the parental lines. Interestingly, only five PAV genes were identified with significantly differential expression in the hybrid. It suggests that the expression of only a part of PAV genes was consistent with the complementation hypothesis. In conclusion, it is unlikely that heterosis is the result of any single mechanism [58,61].

EXPERIMENTAL SECTION

Plant Growth and RNA Isolation

The hybrid corn combination, B73 × Mo17 (F_1 and its parental lines, B73 and Mo17), was originally obtained from Thomas Lübberstedt

(Iowa State University, Ames, IA, USA). The materials were offered for high-throughput sequencing and quantitative real time PCR (qRT-PCR) analysis. The inbred lines were cultivated at the experimental station of Sichuan Agricultural University, Chengdu, for seed propagation of the inbreds, and for production of B73 × Mo17 hybrid seed. Kernels of the combination were grown in environmentally controlled growth chambers that provided 15 h of light (25 °C) and 9 h of dark (20 °C) as described previously. Light intensity was ≈650–800 μmol·m^{-2}·s^{-1}. Ears of 10 random healthy plants at early flower differentiation stages were manually collected as a pool for each genotype as described previously [62]. In brief, ears at stage 3 (three stamen primordia and one pistil primordium can be observed) were collected according to the plant features (number of visible leaves, leaf age index, number of unfolded and folded leaves) combined with microscopic observation. At stage 3, the number of visible leaves, leaf age index, and the number of unfolded and folded leaves are 19, 65%, 13.7 and 5, respectively. Morphological observations of ears at stage 3 in two inbred lines and their F_1 were almost the same as that in another inbred line X178 reported previously [62]. After separately grinding meristems in liquid nitrogen, RNA were extracted for constructing three digital gene expression libraries, and quantitative real-time PCR validation from ≈5 g of frozen tissue by using TRIzol reagent (Invitrogen, Carlsbad, CA, USA) according to the manufacturer's instructions.

Digital Gene Expression Library Preparation and Sequencing

Tag library preparation for the three genotypes (B73, Mo17 and their F_1) was performed in parallel by using the Illumina gene expression sample preparation kit as described previously [63]. An extract of 6 μg of total RNA was obtained and treated with Oligo (dT) magnetic bead adsorption to purify mRNA. Oligo (dT) was then used as a primer to synthesize the first- and second-strand cDNA. The 5› ends of tags were generated by two endonucleases *Nla*III or *Dpn*II. The bead-bound cDNA was subsequently digested with restriction enzyme *Nla*III, which recognizes and cuts off the CATG sites. The fragments apart from the 3′ cDNA fragments connected to Oligo (dT) beads were washed away and the Illumina adaptor 1 was ligated to the sticky 5′ end of the digested bead-bound cDNA fragments. The junction of Illumina adaptor 1 and CATG site is the recognition site of *Mme*I, which is a type of Endonuclease with separate recognition and digestion sites. It cuts 17 bp downstream of the CATG site, producing tags with adaptor 1. After removing 3′ fragments with magnetic beads by precipitation, Illumina adaptor 2 was ligated to the 3′ ends of tags, acquiring tags with different adaptors at both ends to form a tag library. After 15 cycles of linear PCR amplification, 95 bp fragments were purified by 6% TBE PAGE Gel electrophoresis. After denaturation, the single-chain molecules were fixed onto the Illumina Sequencing flow cell. Each molecule grows into a single-molecule cluster sequencing template through *in situ* amplification. The four types of nucleotides were labeled by four colors, and added to perform sequencing by synthesis (SBS) [64]. Each tunnel will generate millions of raw reads with sequencing length of 35 bp.

Quantitative RT-PCR and Gene Expression Analysis

In order to verify a sample of genes that exhibited statistically significant differential expression in the analysis of DGE data, we used quantitative real time PCR analysis. The RNA samples used for the qRT-PCR assays were the same as for the DGE experiments from 10 biological replicates. First, 1 μg of RNA was treated with RNase-free DNase (Promega, Madison, WI, USA), and cDNA was synthesized with PrimeScript RT reagent kit (TaKaRa, Tokyo, Japan). Then, qRT-PCR of 30 differentially expressed genes (Table S16), which were involved in arginine and proline metabolism, pyruvate metabolism or other genes, were performed using the SYBR *PremixExTaq*™ protocol (TaKaRa, Tokyo, Japan) on an Applied Biosystems 7500 Real-Time PCR System (Applied Biosystems, Foster City, CA, USA). For each sample, measurements were performed in triplicate, and the average cycle thresholds (C_t) were used to determine fold-change. 18S rRNA was employed as an endogenous control. The results were calculated using the $2^{-\Delta\Delta Ct}$ method [65].

Analysis and Mapping of Digital Gene Expression Tags

Raw sequencing image data were transformed by base calling into sequences. These raw data reads were stored in FASTQ format, and their analysis conducted as described by Qin *et al.* [63]. In brief, prior to mapping to the reference database, all sequences were filtered to trim the 3› adaptor sequence, filter empty tags (reads with only 3› adaptor sequences but no tags) and low-quality tags containing Ns, and remove tags which are too long or too short. A virtual library containing all possible CATG + 17 base-length sequences of the maize genome database (AGPv2, release 5b.60) [43] was utilized. All clean tags were mapped to the reference sequences and a mismatch of only 1 bp was considered. Clean tags that were mapped to the maize genome reference sequences from multiple genes were filtered. The remaining clean tags were designed as unambiguous clean tags. The expression level of each gene was estimated by the frequency of clean tags and then normalized to TPM (number of transcripts per million clean tags) [66], which is a standard method and extensively used in DGE analysis [67]. KOG functional classification, Gene Ontology (GO), pathway annotation and enrichment analyses were based on the NCBI COG (http://www.ncbi.nlm.nih.gov/COG) [68], Gene Ontology Database (http://www.geneontology.org/) [69] and KEGG pathway (http://www.genome.jp/kegg/) [70], respectively. When we investigate pathways in which DGs were involved and enriched, *q*-value was used for aided identification according to the previous description [47].

Identification of Differentially Expressed Genes

To examine differential expression across samples (B73, Mo17 and their hybrid), the number of raw clean tags in each library was normalized to TPM to obtain normalized gene expression levels. Detection of different tags across samples were performed as previously described [71]. The false discovery rate (5%) is controlled by the Benjamini and Hochberg's procedure [72]. After multiple testing between pairwise comparisons, we use "FDR ≤ 0.001 and the absolute value of $\log_2$Ratio ≥ 1" as the threshold to judge the significance of gene expression difference. More stringent criteria with smaller FDR and bigger fold-change values were used to identify different genes.

In the present study, the same strategy was performed in a linear-in-genotype contrast when F_1 genotype was compared to the two parental lines as described by Zhang *et al.* [16]. The genes with "FDR $\leq$ 0.001 and the absolute value of $\log_2$Ratio $\geq$ 1" were regarded as non-additivity, when the genes with "FDR > 0.001 and the absolute value of $\log_2$Ratio < 1" were regarded as not statistically significantly different from additivity. To classify the genes further, the high-parent dominant genes and low-parent dominant genes were identified from the non-additive group based on the criterion, that the F_1 genotype was significantly different from one parent and not significantly different from another parent. From the non-additive group, the expression of genes was identified as over- or under-dominant, when expression in the F_1 genotype was significantly higher or lower than in both inbred parents, respectively.

CONCLUSIONS

Our analysis revealed 17,128 genes expressed in these three genotypes and 22,789 genes expressed collectively in the present study. Approximately 38% of the genes were differentially expressed in early maize ear inflorescences from heterotic cross, including many transcription factor genes and some presence/absence variations (PAVs) genes, and exhibited multiple modes of gene action. Additionally, a significant number of genes were expressed in only one inbred line or absent in both inbred lines. Comparison of the differences of modes of gene action between previous studies and the present study revealed only small number of different genes had the same modes of gene action in both maize seedlings and ear inflorescences, it might of be an indication that, in different tissues or developmental stages, different global expression patterns might prevail, which might nevertheless be related to heterosis. Our results support the hypothesis that multiple molecular mechanisms (dominance and overdominance modes) contribute to heterosis.

ACKNOWLEDGMENTS

This work was supported by National modern agricultural industry technology system, the Major Project of China on New varieties of GMO Cultivation (2011ZX08003-003), National Natural Science Foundation of China (30201639), and the national Hi-Tec program of China (2012AA10307).

AUTHOR CONTRIBUTIONS

H.D., G.P., C.Q. and Z.Z. designed the study; C.Q., X.L., L.L., J.G., H.L., Y.S. and M.Z. performed the analyses; C.Q and Z.C. conducted qRT-PCR; H.D., C.Q. and T.L. drafted the manuscript. All of the authors critically revised and provided final approval of this manuscript.

CONFLICTS OF INTEREST

The authors declare no conflict of interest.

REFERENCES

1. Shull G.H. The composition of a field of maize. J. Hered. 1908;4:296–301. doi: 10.1093/jhered/os-4.1.296.
2. Bruce A.B. The Mendelian theory of heredity and the augmentation of vigor. Science. 1910;32:627.
3. Jones D.F. Dominance of linked factors as a means of accounting for heterosis. Genetics. 1917;6:466–479.
4. Davenport C.B. Degeneration, albinism and inbreeding. Science. 1908;28:454.
5. East E.M. Heterosis. Genetics. 1936;21:375–397.
6. Birchler J.A., Auger D.L., Riddle N.C. In search of the molecular basis of heterosis. Plant Cell.2003;15:2236–2239. doi: 10.1105/tpc.151030.
7. Yu S.B., Li J.X., Xu C.G., Tan Y.F., Gao Y.J., Li X.H., Zhang Q.F.,

Saghai Maroof M.A. Importance of epistasis as the genetic basis of heterosis in an elite rice hybrid. Proc. Natl. Acad. Sci. USA. 1997;99:9226–9231.

8. Li Z.K., Luo L.J., Mei H.W., Wang D.L., Shu Q.Y., Tabien R., Zhong D.B., Ying C.S., Stansel J.W., Khush G.S., et al. Overdominant epistatic loci are the primary genetic basis of inbreeding depression and heterosis in rice. I. Biomass and grain yield. Genetics. 2001;158:1737–1753.

9. Xiao J., Li J., Yuan L., Tanksley S.D. Dominance is the major genetic basis of heterosis in rice as revealed by QTL analysis using molecular markers. Genetics. 1995;140:745–754.

10. Stuber C.W., Lincoln S.E., Wolff D.W., Helentjaris T., Lander E.S. Identification of genetic factors contributing to heterosis in a hybrid from two elite maize inbred lines using molecular markers. Genetics.1992;132:823–839.

11. Auger D.L., Gray A.D., Ream T.S., Kato A., Coe E.H., Jr., Birchler J.A. Nonadditive gene expression in diploid and triploid hybrids of maize. Genetics. 2005;169:389–397.

12. Hoecker N., Keller B., Muthreich N., Chollet D., Descombes P., Piepho H.P., Hochholdinger F. Comparison of maize (*Zea mays* L.) F_1-hybrid and parental inbred line primary root transcriptomes suggests organ-specific patterns of nonadditive gene expression and conserved expression trends. Genetics.2008;179:1275–1283.

13. Song R., Messing J. Gene expression of a gene family in maize based on noncollinear haplotypes. Proc. Natl. Acad. Sci. USA. 2003;100:9055–9060.

14. Stupar R.M., Springer N.M. *cis*-Transcriptional variation in maize inbred lines B73 and Mo17 leads to additive expression patterns in the F_1 hybrid. Genetics. 2006;173:2199–2210. doi: 10.1534/genetics.106.060699.

15. Vuylsteke M., van Eeuwijk F., van Hummelen P., Kuiper M., Zabeau M. Genetic analysis of variation in gene expression in *Arabidopsis thaliana*. Genetics. 2005;171:1267–1275.

16. Zhang H.Y., He H., Chen L.B., Li L., Liang M.Z., Wang X.F., Liu X.G., He G.M., Chen R.S., Ma L.G., et al. A genome-wide transcription analysis reveals a close correlation of promoter

INDEL polymorphism and heterotic gene expression in rice hybrids. Mol. Plant. 2008;1:720–731. doi: 10.1093/mp/ssn022.

17. Guo M., Yang S., Rupe M., Hu B., Bickel D., Arthur L., Smith O. Genome-wide allele-specific expression analysis using Massively Parallel Signature Sequencing (MPSS™) Reveals *cis*- and *trans*-effects on gene expression in maize hybrid meristem tissue. Plant Mol. Biol. 2008;66:551–563.
18. Chen X., Li M., Shi J., Fu D., Qian W., Zou J., Zhang C., Meng J. Gene expression profiles associated with intersubgenomic heterosis in *Brassica napus*. Theor. Appl. Genet. 2008;117:1031–1040. doi: 10.1007/s00122-008-0842-z.
19. Zhuang Y., Adams K.L. Extensive allelic variation in gene expression in populus F_1 hybrids. Genetics.2007;177:1987–1996. doi: 10.1534/genetics.107.080325.
20. Jahnke S., Sarholz B., Thiemann A., Kühr V., Gutiérrez-Marcos J., Geiger H., Piepho H.-P., Scholten S. Heterosis in early seed development: A comparative study of F_1 embryo and endosperm tissues 6 days after fertilization. Theor. Appl. Genet. 2010;120:389–400. doi: 10.1007/s00122-009-1207-y.
21. Pea G., Ferron S., Gianfranceschi L., Krajewski P., Enrico Pe M. Gene expression non-additivity in immature ears of a heterotic F_1 maize hybrid. Plant Sci. 2008;174:17–24.
22. Riddle N., Jiang H., An L., Doerge R., Birchler J. Gene expression analysis at the intersection of ploidy and hybridity in maize. Theor. Appl. Genet. 2010;120:341–353. doi: 10.1007/s00122-009-1113-3.
23. Guo M., Rupe M.A., Yang X., Crasta O., Zinselmeier C., Smith O.S., Bowen B. Genome-wide transcript analysis of maize hybrids: Allelic additive gene expression and yield heterosis. Theor. Appl. Genet.2006;113:831–845. doi: 10.1007/s00122-006-0335-x.
24. Paschold A., Marcon C., Hoecker N., Hochholdinger F. Molecular dissection of heterosis manifestation during early maize root development. Theor. Appl. Genet. 2010;120:383–388. doi: 10.1007/s00122-009-1082-6.
25. Meyer S., Pospisil H., Scholten S. Heterosis associated gene expression in maize embryos 6 days after fertilization exhibits additive, dominant and overdominant pattern. Plant Mol. Biol. 2007;63:381–391. doi: 10.1007/s11103-006-9095-x.

26. Stupar R.M., Gardiner J.M., Oldre A.G., Haun W.J., Chandler V.L., Springer N.M. Gene expression analyses in maize inbreds and hybrids with varying levels of heterosis. BMC Plant Biol. 2008;8:33.
27. Uzarowska A., Keller B., Piepho H.P., Schwarz G., Ingvardsen C., Wenzel G., Lubberstedt T. Comparative expression profiling in meristems of inbred-hybrid triplets of maize based on morphological investigations of heterosis for plant height. Plant Mol. Biol. 2007;63:21–34.
28. Sun Q., Wu L., Ni Z., Meng F., Wang Z., Lin Z. Differential gene expression patterns in leaves between hybrids and their parental inbreds are correlated with heterosis in a wheat diallel cross. Plant Sci.2004;166:651–657. doi: 10.1016/j.plantsci.2003.10.033.
29. Swanson-Wagner R.A., DeCook R., Jia Y., Bancroft T., Ji T., Zhao X., Nettleton D., Schnable P.S. Paternal dominance of *trans*-eQTL influences gene expression patterns in maize hybrids. Science.2009;326:1118–1120.
30. Wei G., Tao Y., Liu G., Chen C., Luo R., Xia H., Gan Q., Zeng H., Lu Z., Han Y., et al. A transcriptomic analysis of superhybrid rice LYP9 and its parents. Proc. Natl. Acad. Sci. USA.2009;106:7695–7701.
31. Swanson-Wagner R.A., Jia Y., DeCook R., Borsuk L.A., Nettleton D., Schnable P.S. All possible modes of gene action are observed in a global comparison of gene expression in a maize F_1 hybrid and its inbred parents. Proc. Natl. Acad. Sci. USA. 2006;103:6805–6810.
32. Song G.S., Zhai H.L., Peng Y.G., Zhang L., Wei G., Chen X.Y., Xiao Y.G., Wang L., Chen Y.J., Wu B., et al. Comparative transcriptional profiling and preliminary study on heterosis mechanism of super-hybrid rice. Mol. Plant. 2010;3:1012–1025. doi: 10.1093/mp/ssq046.
33. Chen Z.J. Genetic and epigenetic mechanisms for gene expression and phenotypic variation in plant polyploids. Annu. Rev. Plant Biol. 2007;58:377–406. doi: 10.1146/annurev.arplant.58.032806.103835.
34. Ha M., Lu J., Tian L., Ramachandran V., Kasschau K.D., Chapman E.J., Carrington J.C., Chen X., Wang X.J., Chen Z.J.

Small RNAs serve as a genetic buffer against genomic shock in *Arabidopsis*interspecific hybrids and allopolyploids. Proc. Natl. Acad. Sci. USA. 2009;106:17835–17840.

35. He G., Zhu X., Elling A.A., Chen L., Wang X., Guo L., Liang M., He H., Zhang H., Chen F., et al. Global epigenetic and transcriptional trends among two rice subspecies and their reciprocal hybrids. Plant Cell. 2010;22:17–33.
36. Liu H., Qin C., Chen Z., Zuo T., Yang X., Zhou H., Xu M., Cao S., Shen Y., Lin H., et al. Identification of miRNAs and their target genes in developing maize ears by combined small RNA and degradome sequencing. BMC Genomics. 2014;15:25.
37. Hollick J.B., Chandler V.L. Epigenetic allelic states of a maize transcriptional regulatory locus exhibit overdominant gene action. Genetics. 1998;150:891–897.
38. Zhao X., Chai Y., Liu B. Epigenetic inheritance and variation of DNA methylation level and pattern in maize intra-specific hybrids. Plant Sci. 2007;172:930–938. doi: 10.1016/j.plantsci.2007.01.002.
39. Ni Z., Kim E.D., Ha M., Lackey E., Liu J., Zhang Y., Sun Q., Chen Z.J. Altered circadian rhythms regulate growth vigour in hybrids and allopolyploids. Nature. 2009;457:327–331.
40. Wang X., Cao H., Zhang D., Li B., He Y., Li J., Wang S. Relationship between differential gene expression and heterosis during ear development in maize (*zea mays* L.) J. Genet. Genomics. 2007;34:160–170. doi: 10.1016/S1673-8527(07)60017-4.
41. Li R., Yu C., Li Y., Lam T.W., Yiu S.M., Kristiansen K., Wang J. SOAP2: An improved ultrafast tool for short read alignment. Bioinformatics. 2009;25:1966–1967.
42. Li P., Ponnala L., Gandotra N., Wang L., Si Y., Tausta S.L., Kebrom T.H., Provart N., Patel R., Myers C.R., et al. The developmental dynamics of the maize leaf transcriptome. Nat. Genet. 2010;42:1060–1067. doi: 10.1038/ng.703.
43. Schnable P.S., Ware D., Fulton R.S., Stein J.C., Wei F., Pasternak S., Liang C., Zhang J., Fulton L., Graves T.A., et al. The B73 maize genome: Complexity, diversity, and dynamics. Science. 2009;326:1112–1115. doi: 10.1126/science.1178534.

44. Ashburner M., Ball C.A., Blake J.A., Botstein D., Butler H., Cherry J.M., Davis A.P., Dolinski K., Dwight S.S., Eppig J.T., et al. Gene ontology: Tool for the unification of biology—The gene ontology consortium. Nat. Genet. 2000;25:25–29.

45. Carbon S., Ireland A., Mungall C.J., Shu S., Marshall B., Lewis S. AmiGO: Online access to ontology and annotation data. Bioinformatics. 2009;25:288–289. doi: 10.1093/bioinformatics/btn615.

46. Xie C., Mao X., Huang J., Ding Y., Wu J., Dong S., Kong L., Gao G., Li C.-Y., Wei L. KOBAS 2.0: A web server for annotation and identification of enriched pathways and diseases. Nucleic Acids Res.2011;39:W316–W322.

47. Storey J.D., Tibshirani R. Statistical significance for genomewide studies. Proc. Natl. Acad. Sci. USA.2003;100:9440–9445. doi: 10.1073/pnas.1530509100.

48. Mao X., Cai T., Olyarchuk J.G., Wei L. Automated genome annotation and pathway identification using the KEGG Orthology (KO) as a controlled vocabulary. Bioinformatics. 2005;21:3787–3793. doi: 10.1093/bioinformatics/bti430.

49. Wu J., Mao X., Cai T., Luo J., Wei L. KOBAS server: A web-based platform for automated annotation and pathway identification. Nucleic Acids Res. 2006;34:W720–W724. doi: 10.1093/nar/gkl167.

50. Vilella A.J., Severin J., Ureta-Vidal A., Heng L., Durbin R., Birney E. EnsemblCompara GeneTrees: Complete, duplication-aware phylogenetic trees in vertebrates. Genome Res. 2009;19:327–335.

51. Liang C., Jaiswal P., Hebbard C., Avraham S., Buckler E.S., Casstevens T., Hurwitz B., McCouch S., Ni J., Pujar A., et al. Gramene: A growing plant comparative genomics resource. Nucleic Acids Res.2008;36: D947–D953.

52. Eveland A.L., Satoh-Nagasawa N., Goldshmidt A., Meyer S., Beatty M., Sakai H., Ware D., Jackson D. Digital gene expression signatures for maize development. Plant Physiol. 2010;154:1024–1039. doi: 10.1104/pp.110.159673.

53. Fu H., Dooner H.K. Intraspecific violation of genetic colinearity and its implications in maize. Proc. Natl. Acad. Sci. USA. 2002;99:9573–9578.

54. Lai J., Li R., Xu X., Jin W., Xu M., Zhao H., Xiang Z., Song W., Ying K., Zhang M., et al. Genome-wide patterns of genetic variation among elite maize inbred lines. Nat. Genet. 2010;42:1027–1030.

55. Shen Y., Zhang Y., Chen J., Lin H., Zhao M., Peng H., Liu L., Yuan G., Zhang S., Zhang Z., et al. Genome expression profile analysis reveals important transcripts in maize roots responding to the stress of heavy metal Pb. Physiol. Plant. 2012;147:270–282.

56. Xiao S., Jia J., Mo D., Wang Q., Qin L., He Z., Zhao X., Huang Y., Li A., Yu J., et al. Understanding PRRSV infection in porcine lung based on genome-wide transcriptome response identified by deep sequencing. PLoS One. 2010;5:e11377.

57. Gibson G., Riley-Berger R., Harshman L., Kopp A., Vacha S., Nuzhdin S., Wayne M. Extensive sex-specific nonadditivity of gene expression in *Drosophila melanogaster*. Genetics. 2004;167:1791–1799. doi: 10.1534/genetics.104.026583.

58. Charlesworth D., Willis J.H. The genetics of inbreeding depression. Nat. Rev. Genet. 2009;10:783–796. doi: 10.1038/nrg2664.

59. Guo M., Rupe M.A., Zinselmeier C., Habben J., Bowen B.A., Smith O.S. Allelic variation of gene expression in maize hybrids. Plant Cell. 2004;16:1707–1716.

60. Birchler J.A., Riddle N.C., Auger D.L., Veitia R.A. Dosage balance in gene regulation: Biological implications. Trends Genet. 2005;21:219–226. doi: 10.1016/j.tig.2005.02.010.

61. Springer N.M., Stupar R.M. Allelic variation and heterosis in maize: How do two halves make more than a whole? Genome Res. 2007;17:264–275. doi: 10.1101/gr.5347007.

62. Zhu Y., Fu J., Zhang J., Liu T., Jia Z., Wang J., Jin Y., Lian Y., Wang M., Zheng J., et al. Genome-wide analysis of gene expression profiles during ear development of maize. Plant Mol. Biol. 2009;70:63–77. doi: 10.1007/s11103-009-9457-2.

63. Qin Y.F., Fang H.M., Tian Q.N., Bao Z.X., Lu P., Zhao J.M., Mai J., Zhu Z.Y., Shu L.L., Zhao L., et al. Transcriptome profiling and digital gene expression by deep-sequencing in normal/regenerative tissues of planarian *Dugesia japonica*. Genomics. 2011;97:364–371. doi: 10.1016/j.ygeno.2011.02.002.

64. Mardis E.R. The impact of next-generation sequencing technology on genetics. Trends Genet.2008;24:133–141. doi: 10.1016/j.tig.2007.12.007.

65. Livak K.J., Schmittgen T.D. Analysis of relative gene expression data using real-time quantitative PCR and the $2^{-\Delta\Delta Ct}$ method. Methods. 2001;25:402–408.

66. C't Hoen P.A., Ariyurek Y., Thygesen H.H., Vreugdenhil E., Vossen R.H., de Menezes R.X., Boer J.M., van Ommen G.J.B., den Dunnen J.T. Deep sequencing-based expression analysis shows major advances in robustness, resolution and inter-lab portability over five microarray platforms. Nucleic Acids Res. 2008;36:e141. doi: 10.1093/nar/gkn705.

67. Morrissy A.S., Morin R.D., Delaney A., Zeng T., McDonald H., Jones S., Zhao Y., Hirst M., Marra M.A. Next-generation tag sequencing for cancer gene expression profiling. Genome Res. 2009;19:1825–1835. doi: 10.1101/gr.094482.109.

68. Tatusov R.L., Fedorova N.D., Jackson J.D., Jacobs A.R., Kiryutin B., Koonin E.V., Krylov D.M., Mazumder R., Mekhedov S.L., Nikolskaya A.N., et al. The COG database: An updated version includes eukaryotes. BMC Bioinform. 2003;4:41.

69. Consortium T.G.O. The gene ontology project in 2008. Nucleic Acids Res. 2008;36:D440–D444. doi: 10.1093/nar/gkm883.

70. Kanehisa M. "In Silico"Simulation of Biological Processes. John Wiley & Sons; Manhattan, NY, USA: 2008. The KEGG Database; pp. 91–103.

71. Marioni J.C., Mason C.E., Mane S.M., Stephens M., Gilad Y. RNA-seq: An assessment of technical reproducibility and comparison with gene expression arrays. Genome Res. 2008;18:1509–1517. doi: 10.1101/gr.079558.108.

72. Benjamini Y., Hochberg Y. Controlling the false discovery rate: A practical and powerful approach to multiple testing. J. R. Stat. Soc. B. 1995;57:289–300.

Chapter 6

Improved Nutritive Quality and Salt Resistance in Transgenic Maize by Simultaneously Overexpression of a Natural Lysine-Rich Protein Gene, Sbglr, and An Erf Transcription Factor Gene, Tsrf1

Meizhen Wang,[1,2]; Chen Liu,[1] Shixue Li,1Dengyun Zhu,[1]Qian Zhao,[1]and Jingjuan Yu[1]

[1]State Key Laboratory of Agrobiotechnology, College of Biological Sciences, China Agricultural University, No. 2 Yuanmingyuan West Road, Haidian District, Beijing 100193, China;

[2]Institute of Medicinal Plant Development, Chinese Academy of

Medical Sciences & Peking Union Medical College, No. 151, Malianwa North Road, Haidian District, Beijing 100193, China;

ABSTRACT

Maize (*Zea mays* L.), as one of the most important crops in the world, is deficient in lysine and tryptophan. Environmental conditions greatly impact plant growth, development and productivity. In this study, we used particle bombardment mediated co-transformation to obtain marker-free transgenic maize inbred X178 lines harboring a lysine-rich protein gene *SBgLR* from potato and an ethylene responsive factor (ERF) transcription factor gene, *TSRF1*, from tomato. Both of the target genes were successfully expressed and showed various expression levels in different transgenic lines. Analysis showed that the protein and lysine content in T_1 transgenic maize seeds increased significantly. Compared to non-transformed maize, the protein and lysine content increased by 7.7% to 24.38% and 8.70% to 30.43%, respectively. Moreover, transgenic maize exhibited more tolerance to salt stress. When treated with 200 mM NaCl for 48 h, both non-transformed and transgenic plant leaves displayed wilting and losing green symptoms and dramatic increase of the free proline contents. However, the degree of control seedlings was much more serious than that of transgenic lines and much more increases of the free proline contents in the transgenic lines than that in the control seedlings were observed. Meanwhile, lower extent decreases of the chlorophyll contents were detected in the transgenic seedlings. Quantitative RT-PCR was performed to analyze the expression of ten stress-related genes, including stress responsive transcription factor genes, *ZmMYB59* and *ZmMYC1*, proline synthesis related genes, *ZmP5CS1* and *ZmP5CS2*, photosynthesis-related genes, *ZmELIP*, *ZmPSI-N*,*ZmOEE*, *Zmrbcs* and *ZmPLAS*, and one ABA biosynthesis related gene, *ZmSDR*. The results showed that with the exception of *ZmP5CS1* and *ZmP5CS2* in line 9–10 and 19–11, *ZmMYC1* in line 19–11 and*ZmSDR* in line 19–11, the expression of other stress-related genes were inhibited in transgenic lines under normal conditions. After salt treatment, the expressions of the ten stress-related genes were significantly induced in both wild-type (WT) and transgenic lines. However, compared to WT, the increases of *ZmP5CS1*in all these three transgenic lines and *ZmP5CS2* in line

9–10 were less than WT plants. This study provides an effective approach of maize genetic engineering for improved nutritive quality and salt tolerance.

INTRODUCTION

Maize (*Zea mays* L.), also known as corn, is one of the most widely cultivated crops in the world. It is mainly used as human food, livestock feed and industrial raw material. However, malnutrition is common in the countries, where corn is the sole or primary food source, due to the deficiency of essential amino acids, like lysine and tryptophan [1]. Actually, major efforts have been made to identify high-lysine corn varieties by genetic approaches since the mid-twentieth century. *Opaque 2* (*o2*) is a natural high-lysine mutant, which has twice of the lysine content than in the normal maize [2]. However, several drawbacks were observed in this mutant, such as soft chalky endosperm [2]. Even though quality protein maize (QPM) has been produced by backcrossing of *o2* modifier genes into *o2* mutant [3], it is a lengthy progress (~30 years) that limits the spread of QPM. Genetic engineering technologies are also utilized aiming to increase lysine content by zein reduction [4–6]. Unfortunately, all these transgenic plants exhibit opaque kernel phenotypes, which were very similar to that of *o2* mutant [4–6]. Moreover, increase of lysine content in corn grain has been achieved by manipulation of both lysine biosynthesis and metabolic pathways [7,8], including the commercialized high-lysine maize variety LY038 [9]. However, only free lysine content is significantly increased in these transgenic plants [8].

Recently, seed-specific expression of natural proteins with high-lysine concentration has been proved to be an effective approach to increase the lysine content of corn grain. Among them, milk proteins are attractive choice due to their balanced amino acid composition and good digestibility. As reported, lysine content is enhanced obviously in maize endosperm when expressing the milk protein, α-lactalbumin [10,11], whereas total protein content is not significantly different from negative kernels [11]. Liu *et al.* (1997) cloned a pollen-specific cDNA encoding a lysine-rich protein SB401 from potato [12]. In our previous studies, seed-specific expression of *SB401* increases lysine content from 16.1% to 54.8%, and meanwhile, the total protein content is increased from 11.6% to 39.0% in T_1 transgenic maize seeds, compared with the non-transgenic lines [13]. Thereafter, the natural lysine-rich protein gene *SBgLR*, homologous to *SB401*, was isolated from potato and transformed into maize, resulting in greater lysine and protein contents increase in transgenic maize [14]. It provides an effective way for maize quality improvement.

Environmental conditions, such as drought, salinity, extreme temperatures and pathogen attack, greatly affect plant growth, development and productivity. To survive under these conditions, plants have developed a series of defense responsive pathways [15]. In this signaling network, transcription factors play important roles through regulating expression of stress-related genes [16,17]. Ethylene responsive factor (ERF) proteins are a class of plant specific transcription factors, which have been reported to modulate plant development and multiple stress responses [18–22]. It was found that ERF proteins could function not only in plant pathogen resistance by interacting with GCC-box (AGCCGCC) [23–25], but also in abiotic resistance through interacting with the DRE element (CCGAC) [18,21,26–28]. TSRF1, an ERF protein from tomato, can activate the expression of PR genes by binding to the GCC-box and positively regulates pathogen resistance in transgenic tomato and tobacco [29,30]. ABA enhances the binding of TSRF1 to GCC-box and modifies the pathogen resistance of transgenic tobacco [31]. In addition, TSRF1 negatively regulates the osmotic response in transgenic tobacco [30], but improves the osmotic and drought tolerance in transgenic rice, indicating different mechanisms of TSRF1 in dicots and monocots [32]. All these results suggest that genetic engineering of ERF transcription factor genes is an effective way to improve plant stress tolerance [33].

Genetically modified (GM) crops are the fast-adopted technology in agriculture [34]. It has increased to 29 countries growing GM crops in recent years [35]. However, intense debates have been addressed regarding the safety of GM crops to the environment and customers. Selectable marker genes, like herbicide or antibiotic resistance genes, are supposed to be dangerous, although there is no strong scientific evidence [36]. By far, various strategies to eliminate the selectable marker genes have been developed and successfully used, such as co-transformation, site-specific recombination, multi-auto transformation vector, transposition system and homologous recombination (see more in review [35]). For *Agrobacterium*-mediated co-transformation, there are three methods: (i) two different vectors carried by different *Agrobacterium* strains [37]; (ii) two different vectors in the same *Agrobacterium* strain [38] and (iii) one binary vector with twin T-DNAs [39]. For biolistic bombardment mediated co-transformation, two different plasmids were introduced into the same tissue [40,41]. Among them, co-transformation is widely used due to its simplicity. It can be carried out either by *Agrobacterium*-mediated transformation or by biolistic bombardment. Selectable marker genes and target genes are usually integrated into the separate loci and segregate independently in the subsequent generations [42–44].

In this study, two plasmids containing the target genes, a lysine-rich protein gene, *SBgLR*, and an ERF transcription factor gene, *TSRF1*, and the selectable marker gene, *Hpt*, respectively, were introduced into maize elite inbred X178 embryogenic calli by particle bombardment. Marker-free transgenic maize lines were obtained. Further analysis revealed that both of the target genes were successfully expressed. Further, the lysine and protein content of transgenic maize seeds were detected and the stress tolerance was analyzed by measuring the free proline and chlorophyll contents. The results showed that transgenic lines had not only increased lysine and protein content, but also more tolerance to salt stress. Moreover, ten stress-related gene expression changes were analyzed using quantitative RT-PCR. Marker-free transgenic maize lines with high quality and stress resistance obtained in this research will be good materials for corn breeding.

RESULTS AND DISCUSSION

TSRF1, but Not SBgLR Protein Has Homologs in Maize

To obtain marker-free transgenic maize inbred lines, immature embryogenic calli were transformed by particle bombardment-mediated co-transformation. In this system, two constructs pTSSB and pHpt were used. As is shown in Figure S1, pTSSB contains a potato lysine-rich protein gene, *SBgLR*, under the control of seed-specific promoter, 19Z, and a tomato ERF transcription factor gene, *TSRF1*, under the control of 35S constitutive promoter. The vector pHpt contains a selectable marker gene, *Hpt*, for hygromycin selection. By searching SBgLR protein sequence against the maize reference genome using the tBLASTn program [45,46], no homologs of SBgLR gene were identified (data not shown). However, two homologs for TSRF1 gene were found by similar method (Figure 1). The TSRF1 contains four motifs: an AP2/ERF domain, a CMIX-1 motif, a CMIX-3 motif and a CMIX-4 motif [32]. No ERF proteins containing all the four motifs of TSRF1 have been identified in the japonica rice genome by bioinformatic analysis [47]. To compare the structure of TSRF1 with its homologs in maize, we performed a multiple alignment using Clustal W (http://www.ebi.ac.uk/Tools/clustalw2/index.html) [48]. In contrast to rice, the homologs of maize contain all the four conserved motifs (Figure 1). Therefore, we designed the specific primer pairs in the non-conserved region of *TSRF1* for subsequent PCR and RT-PCR detection of transgenic lines (Table S1).

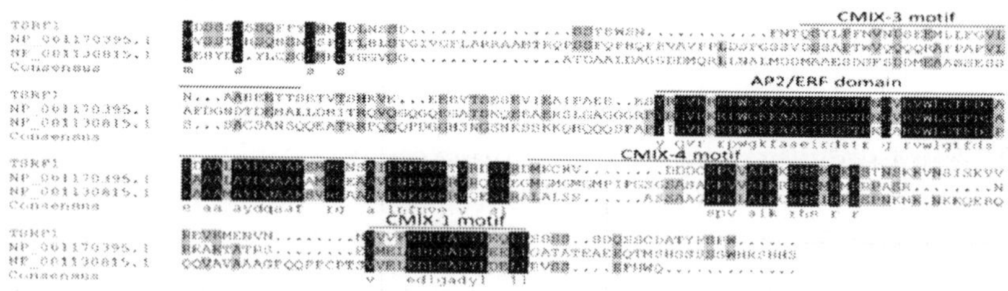

Figure 1: Comparison of the putative motifs of *TSRF1* and its homologs in maize. Motifs of TSRF1 were marked by black lines on the top of the sequences.

Generation of Transgenic Maize Inbred Lines

For co-transformation, two constructs pTSSB and pHpt mixture at a mole ratio of 1.5:1 were co-bombarded into maize embryogenic calli. Different stages of transformation were shown in Figure S2A–F. A total of 114 fertile plants were obtained and self-pollinated for seeds set. In order to confirm the integration of transgenes, we conducted PCR analysis for *SBgLR, TSRF1* and *Hpt* genes. Partial results were shown in Figure S2G,H. Twenty-six transgenic lines positive for all three genes were identified. Transformation efficiency of 1.08% in this study was obtained (Table 1). Both target genes tend to insert into the same loci of one transgenic line. A similar result was reported in maize that an entire 10-member kafirin gene cluster was transformed into maize genome by particle bombardment method without gene silencing [49].

Table 1: Efficiency of particle bombardment-mediated maize co-transformation

Ratio of pTSSB/ pHpt	No. of transformed calli	No. of hygromyc-inresistant events recovered	No. of regenerated plants	PCR analysis	Co-transformation efficiency (%)
				SBgLR **(+)** ***TSRF1*** **(+)** ***Hpt*** **(+)**	
1.5:1	2411	224	114	26	1.08 [a]

[a]Co-transformation efficiency = (No. of all three genes PCR positive plants/No. of calli transformed) × 100%.

Overexpression of SBgLR and TSRF1 in Transgenic Maize

To test whether *SBgLR* and *TSRF1* expressed in transgenic maize, we conducted semi-quantitative RT-PCR using cDNAs from T_1 transgenic maize immature seeds at 22 days after pollination (DAP) and leaves as templates, respectively (Figure 2A,B). The results revealed that both *SBgLR* and *TSRF1* were expressed in the progeny from 16 transgenic lines, and they showed various expression levels among different transgenic lines.

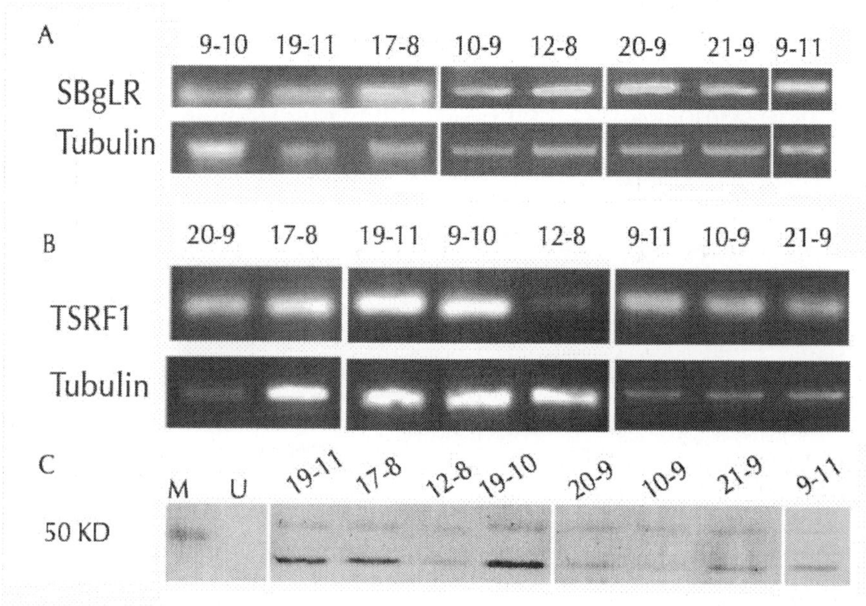

Figure 2: Expression of transgenes in T_1 maize (partial results are shown). (A) Semi-quantification RT-PCR analysis of *SBgLR* in transgenic maize immature seeds (22 DAP); (B) semi-quantification RT-PCR analysis of *TSRF1* in transgenic maize leaves; (C) Western blot analysis of SBgLR in transgenic maize immature seeds (20 DAP). M, protein marker; U, non-transformed maize plants. Numbers on the top of each figure represent the transgenic line number.

Water-soluble proteins were extracted from the immature seeds (20 DAP) of T_1 generation of *SBgLR* and *TSRF1* RT-PCR positive lines. Western blot was further carried out to confirm SBgLR protein accumulation in transgenic maize seeds (Figure 2C). Specific rabbit polyclonal antiserum against SBgLR at 1:400 was used. The predicted molecular mass of the SBgLR was 23 kD, whereas it was much larger (about 50 kD) after SDS/PAGE separation. This discrepancy was also observed in previous Western blot analysis of SBgLR and SB401 in transgenic maize [13,14]. Studies have shown that SB401 is a microtubule associated protein and has a reduced mobility on SDS/PAGE [13,50], which is a common feature of microtubule-associated proteins [51,52]. In addition, it should be noted that two specific bands were observed in Western blot analysis of SBgLR (Figure 2C). The glycosylation site existed in the *N*-terminal of SBgLR may result in the reduction of mobility. However, further confirmation is needed. The transgenic lines with relative high*SBgLR* and *TSRF1* expression levels were selected for subsequent nutritive quality and stress tolerance analysis.

Marker-Free Transgenic Lines Obtained by Segregation in T_2 Generation

In the particle bombardment-mediated co-transformation system, marker gene elimination is dependent on the segregation of the target gene and selectable marker gene in the progeny. To examine whether the selectable marker gene *Hpt* was removed successfully, we performed genomic DNA dot blot to analyze T_2 transgenic lines (Figure 3). Purified PCR fragments of *SBgLR* and *Hpt* were used as probes, respectively. The results indicated that segregation of target genes and *Hpt* occurred in T_2 generation. Two transgenic lines 17 and 15 in R_0 generation contained both target genes and selectable marker gene, but their corresponding T_2generation transgenic lines 17-8-3 and 15-9-1 harbored only target genes as highlighted by arrows. These results indicated that it is a feasible approach to obtain marker-free transgenic lines via particle bombardment mediated co-transformation.

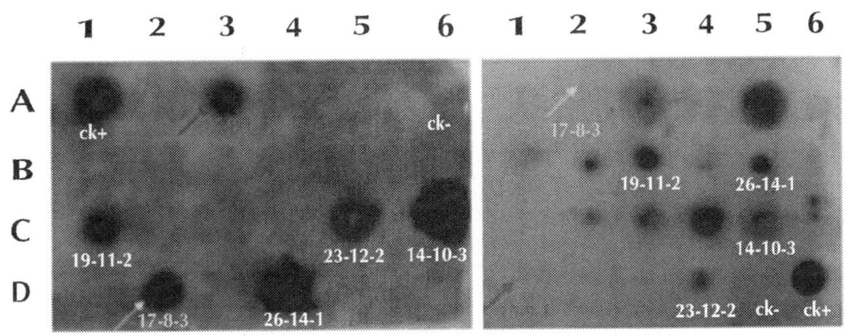

Figure 3: Segregation of target genes and selectable marker gene in T_2 transgenic plants by DNA dot blot analysis. Two segregated transgenic lines are highlighted in red for 15-9-1 and yellow for 17-8-3, respectively. Transgenic lines that have only Hpt are not indicated. For SBgLR (left), A1 represents positive control and A6 represents negative control. For Hpt (right), D6 represents positive control and D5 represents negative control.

SBgLR Enhanced Crude Protein and Lysine Contents in Transgenic Maize

Twenty kernels from each of the self-pollinated transgenic and non-transformed T_1 plants were randomly selected and the lysine and protein contents were analyzed. The results showed that the lysine and protein contents in the seeds were increased at various levels in different transgenic maize lines. Compared to non-transformed maize, the increases of protein and lysine content ranged from 7.7% to 24.38% (Figure 4A) and from 8.70% to 30.43% (Figure 4B), respectively. Line 19–11 showed high SBgLR accumulation level (Figure 2C) and total protein content, but its lysine content was not analyzed, due to limited amount of seeds. Subsequently, we detected the lysine content of T_2 transgenic seeds (Table S2). The result showed that the lysine contents in all these transgenic lines were increased. In addition, it should be noted that the lysine contents in T_2 transgenic lines were increased to a much greater extent than in T_1 generation, indicating much more protein and lysine accumulation, because the *SBgLR* gene became homozygous. However, more evidence is needed. These transgenics will be good materials in quality maize breeding programs.

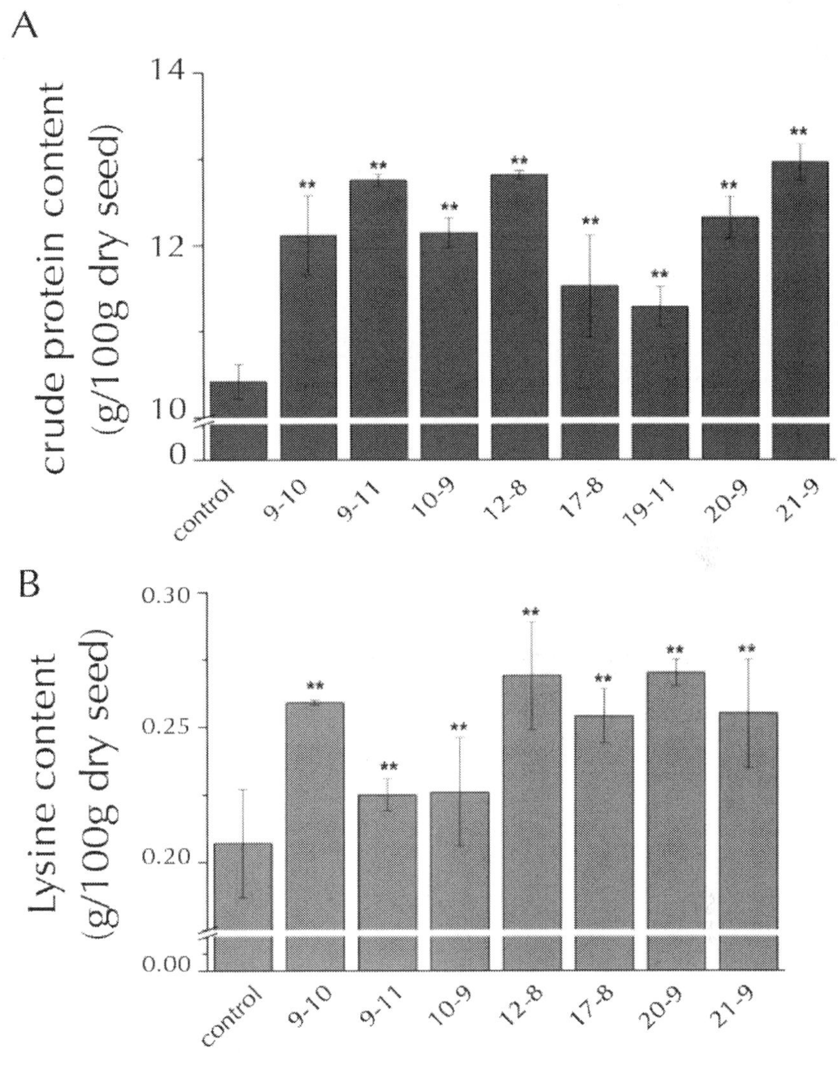

Figure 4: Protein and lysine levels were simultaneously increased in T1 transgenic seeds. (A) Protein content; (B) lysine content. The crude protein and lysine contents are expressed as g/100 g dry seed. Data are averages of triplicates ± standard deviations. Asterisks, **, denote transgenic lines statistically different from Control by Student t-test at $p < 0.01$. The lysine content of line 19–11 was not measured due to limited amount of seeds.

TSRF1 Increased Salt Tolerance in Transgenic Maize

The overexpression of *TSRF1* increases pathogen tolerance in tobacco and tomato [29] and drought resistance in rice [32]. To investigate whether TSRF1 have salt tolerance in transgenic maize, four-leaf stage seedlings of three transgenic lines were used for salt treatment. The result was shown in Figure 5A. Before treatment, these transgenic lines displayed similar growth situation as non-transformed control, suggesting that the expression and insertion of *TSRF1* into maize genome did not influence the growth of transgenic lines. After treated in Hoagland solution containing 200 mM NaCl for 48 h, both non-transformed and transgenic plant leaves displayed wilting and losing green symptoms. However, the degree of control seedlings was much more serious than that of transgenic lines. All the leaves of control seedlings were wilting and rolling. However, the first and second new leaves were still flat and green in transgenic seedlings.

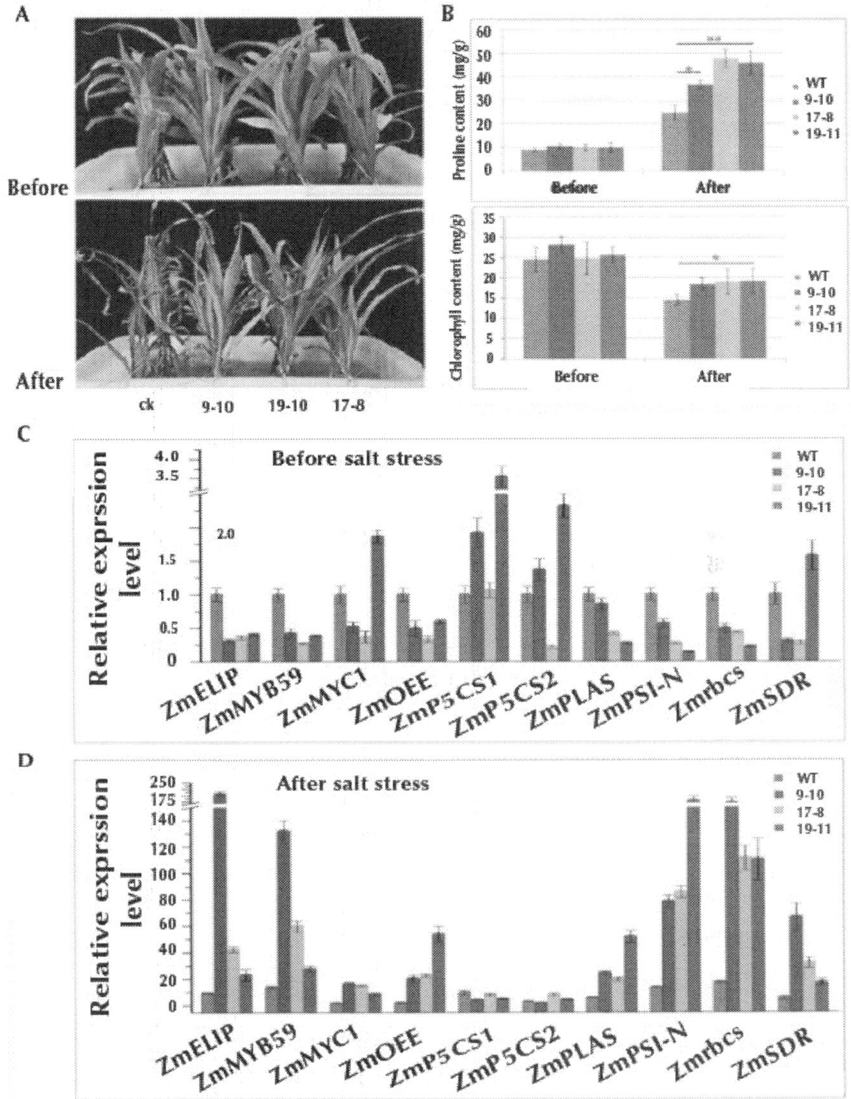

Figure 5: TSRF1 improved salt tolerance in maize. (A) Salt tolerance of TSRF1 transgenic maize. Before, control experiment without drought treatment; after, maize seedlings at four-leaf stage were subjected to 200 mM NaCl stress for 48 h; (B) proline and chlorophyll content in the control and transgenic maize. Data are averages of triplicate ± standard deviations. Asterisks, * and **, denote statistically different from wild-type by Student t-test at $p < 0.05$ and $p < 0.01$, respectively; (C) relative expression of stress-related genes

in wild-type and TSRF1transgenic maize seedlings before salt treatment; (D) relative expression of stress-related genes in wild-type and TSRF1 transgenic maize seedlings after salt treatment. Total RNA extracted from leaves of maize seedlings before and after 200 mM NaCl treatment for 48 h was used, respectively. Tubulin was used as a reference gene. Data are averages of triplicate ± standard deviations.

Proline was considered to be an osmoprotectant to protect plants against the damage resulting from abiotic stress [53]. Plants accumulate high level of proline under stress conditions to prevent the detrimental change. In addition, there is strong evidence that salt stress affects photosynthetic enzymes, chlorophyll and carotenoids in plants [54]. Therefore, we measured the free proline and chlorophyll content change in control and transgenic lines after salt treatment. As it is shown in Figure 5B, the free proline contents in control and transgenic seedlings were at similar level before treatment and increased dramatically after salt treatment. However, the proline content in the three transgenic lines increased at a much larger extent. Similarly, the chlorophyll content in transgenic seedlings decreased at a lower extent after salt treatment, indicating the more tolerance of *TSRF1* overexpressing transgenic maize.

It was previously shown that the constitutive expression of *TSRF1* enhanced rice salt and drought resistance by activating the expression of MYB, MYC, proline synthesis and photosynthesis-related genes [32]. Therefore, we further investigated whether TSRF1 affects the expression of these genes in maize. Ten genes were selected to be analyzed, and the results were shown in Figure 5C,D. The expressions of these genes were found to be different between transgenic lines. This might be due to the different transgenic lines with different target gene insertion sites and the expression levels of *TSRF1*. Under normal conditions, the expression level of *ZmMYC1* in line 19–11, *ZmP5CS1* (Δ^1-pyrroline-5-carboxylate synthetase 1) and *ZmP5CS2* (Δ^1-pyrroline-5-carboxylate synthetase 2) in line 9–10 and 19–11 and *ZmSDR* (short-chain dehydrogenase/reductase) in line 19–11 were increased, while the expression of these four genes in other lines were inhibited. The other six genes, including stress responsive transcription factor genes—*ZmMYB59*—photosynthesis-related genes—*ZmELIP* (low molecular mass early light-inducible protein HV90), *ZmPSI-N* (photosystem I reaction center subunit N), *ZmOEE* (photosystem II oxygen-evolving complex protein), *Zmrbcs* (encoding ribulose 1, 5-bisphosphate carboxylase small subunit) and *ZmPLAS* (encoding plastocyanin)—were inhibited in all these three transgenic lines, which was different from that in rice [32]. This indicated that TSRF1 was probably involved in different regulatory pathways in maize from that in rice. Furthermore, we detected the expression changes of these genes in wild-type and transgenic lines after salt stress. As is shown in Figure 5C, the expression of all these ten genes were increased by 2–16-fold in wild-type after salt treatment, indicating the endogenous regulation of salt responsive genes. Intriguingly, much more fold accumulations of these genes were induced after salt stress in *TSRF1* overexpressing lines. For example, the expression of *Zmrbcs* had increased as high as 171-, 111- and 110-fold in transgenic lines 9–10, 17–8 and 19–11, respectively, compared with wild-type. The sharp induction of photosynthesis-related genes (*ZmELIP*, *Zmrbcs* and *ZmPSI-N*) was consistent with the result of chlorophyll content change in Figure 5B. TSRF1 increases ABA sensitivity in tobacco [31,55] and rice [32]. *SDR*, an ABA biosynthesis related gene, was increased in *TSRF1* overexpressing rice [32]. On the contrary, it was slightly suppressed in line 9–10 and 17–8 under normal condition. After salt treatment, surprisingly, TSRF1 triggered significant *ZmSDR*induction. All these results suggested the complicated interaction of TSRF1 and ABA in plants. The accumulation of transcription factor gene *ZmMYB59* increased much more than *ZmMYC1* in all the transgenic lines, demonstrating the different roles of stress responsive transcription factors in salt stress. They may induce different other downstream salt related genes. These results indicated the complexity mechanism of *TSRF1* that involved in maize salt stress regulatory network.

Moreover, we noted that the expression levels of *ZmP5CS1* and *ZmP5CS2* were not or slightly increased in transgenic lines under salt condition, compared to wild-type, even though high level of proline accumulation was observed (Figure 5B,C). It is reported that the activity of P5CS, a key enzyme representing a rate-limiting step in proline biosynthesis, is regulated through feedback inhibition by proline [56–59]. It is possible that a large amount of proline is synthesized at the early stage of salt stress and maintained this high level throughout the salty condition. Therefore, high activity of P5CS or high level of *P5CS* transcript might be not necessary for high proline content maintaining under salt condition.

EXPERIMENTAL SECTION

Plasmid Construction

Diagrams of two plasmids used in this study were shown in Figure S1. The plasmid pTSRF1 containing the*TSRF1* expression cassette was kindly provided by Prof. Rongfeng Huang, Biotechnology Research Institute, China Academy of Agricultural Sciences (CAAS). The TSRF1 expression cassette was obtained by digesting pTSRF1 with *Bam*HI and *Hin*dIII, blunt-ended with Klenow enzyme (Promega, Madison, WI, USA) and then inserted into the blunt-ended *Hin*dIII site of p19SB, harboring the *SBgLR* expression cassette. The plasmid pHpt was constructed by Wang *et al.* 2006 [44].

Plant Transformation and Detection

Maize inbred line X178, the male parent of the leading commercial cultivar Nongda108 in China, was grown in the test field and was self-pollinated. Embryogenic calli were induced from immature embryos at 10–12 DAP and transformed by particle bombardment, as described by Yu *et al.* 2004 [13].

For transgene detection, genomic DNA was extracted from leaf of transformed plants using SDS method [60], and PCR amplification was performed as follows: 95 °C for 10 min, 30 cycles of 95 °C for 30 s, 54 °C for 30 s and 72 °C for 30 s.

The expression of *SBgLR* and *TSRF1* in transgenic plants was further determined by RT-PCR. Total RNA was extracted from immature seeds (22 DAP) and young leaves, respectively, using TRIzol reagent (Invitrogen, Carlsbad, CA, USA). The first strand of cDNA was synthesized according to the manufacturer's protocol of Reverse Transcription System (Promega, Madison, WI, USA). *Tubulin* was used as the endogenous reference. PCR reactions were carried out for 5 min at 95 °C, followed by 25 cycles of 40 s at 94 °C, 40 s at 54 °C (*SBgLR* and *TSRF1*) or 52 °C (*Tubulin*), 40 s at 72 °C and a final extension of 10 min at 72 °C. Genomic DNA dot blot analysis was used to detect marker-free progeny of T_2 plants. Total genomic DNA was isolated from fresh leaves using the cetyl trimethyl ammonium bromide (CTAB) procedure [61]. A 10 μg aliquot of DNA was spotted onto nylon membrane (Amersham Pharmacia, Arlington Heights, IL, USA). Filters were hybridized using specific purified PCR fragments of *SBgLR* and *Hpt*, respectively, as described for Southern blot by Yu *et al.* 2004 [13]. The primer pair for Hpt gene was the same as reported [14]. The primer pairs for *SBgLR, TSRF1* and *Tubulin* were listed in Table S1.

The lysine-rich protein SBgLR accumulation in transgenic plant seeds was examined by Western blot, as described by Lang *et al.* 2004 [14]. Specific rabbit polyclonal SBgLR antiserum was prepared by the Institute of Genetics and Development Biology, Chinese Academy of Sciences (CAS).

Salt Stress Treatment

We determined the salt tolerance assay of transgenic maize at T_3 generation. Seeds were sowed in nurse soil and grown to four-leaf stage at the condition of 28 °C and 16 h light/8 h dark. Before salt treatment, leaves of one half wide type and transgenic lines were collected and immediately stored in liquid nitrogen, respectively. The other half was used for subsequent free proline and chlorophyll content analysis. For salt tolerance assay, after washing the soil off the roots, the seedlings of wild-type and transgenic lines were treated in Hoagland solution with 200 mM NaCl for 48 h, respectively. As described above, leaves of one-half salt treated wild-type and transgenic seedlings were collected and immediately stored in liquid nitrogen, respectively. The other half was used for subsequent free proline and chlorophyll content analysis.

Detection of Free Proline and Chlorophyll Content

Free proline was extracted from 0.5 g maize leaves (dry weight) using 3% salicylsulfonic acid and then reacted with ninhydrin, followed by determination based on colorimetric method [62].

For chlorophyll a and b content detection, total chlorophyll was extracted from 0.5 g leaves (dry weight), grinded into homogenate in 95% ethanol, and then absorbance at the wavelength 649 and 665 nm was measured under a spectrophotometer (UV2300, Shanghai, China), respectively. The chlorophyll content of each sample was calculated using the equation as follows: $C_a = 13.7A_{665} - 5.76A_{649}$; $C_b = 25.8A_{649} - 7.6A_{665}$; $C_{a+b} = C_a + C_b$.

Gene Expression Analysis by Quantitative Real-Time PCR

Total RNA of maize seedling leaves were extracted using TRIzol reagent (Invitrogen, Carlsbad, CA, USA). After digested with DNase I, 5 μg total RNA was used for cDNA first strand synthesis according to the manufacturer's protocol of Reverse Transcription System (Promega A3500, Madison, WI, USA). For quantitative real-time PCR amplification, 100 ng cDNA was used as template in a 20 μL reaction system, containing 10 μL 2× SYBR Premix Ex Taq II (TaKaRa, Shiga, Japan), 0.5 μM each specific forward and reverse primer. Amplification was carried out on a Bio-Rad Real-Time System CFX96 C1000 Thermal Cycler (Bio-Rad, Hercules, CA, USA) using the following conditions: 95 °C for 30 s, 35 cycles of 95 °C for 10 s, 60 °C for 10 sec and 72 °C for 10 s. Four reference genes, including *Tubulin*, *EF1-α*, *GAPDH* and*Ubiquitin*, were chosen. Expression levels of these four genes were determined as the Ct values [63]. The Ct values were then converted into relative quantities and imported into the geNorm v3,5 software [64] for stability analysis. The result was shown in Figure S3. According to the result, *Tubulin* was selected to be used as the endogenous reference. All primer pairs used for qRT-PCR were listed in Table S1.

Crude Protein and Lysine Content Analysis

Samples were ground meal of 20 mature kernels from individual self-fertilized ears. Due to limited seeds, only 8 T_1 mature kernels for line 19–11 were used. For crude protein content, total nitrogen was first measured based on the principle of Kjeldahl determination under national standard GB2905-82, Beijing Academy of Agriculture and Forestry Science. Then, the nitrogen content was converted to protein content by multiplying a conversion factor of 6.25.

Lysine content was analyzed based on the principle of ninhydrin reaction. Briefly, about 10 mg of defatted powder, 1 mL double distilled water and 2 mL ninhydrin reaction reagent were added into a 30 mL tube, and then the mixture was thoroughly vortexed (approximately 5s). Subsequently, the tube was incubated in boiling water for 20 min, and thereafter, 3 mL 50% ethanol was added to cooled sample, followed by a centrifugation at 13,400× *g* for 10 min. The supernatant was collected, and the absorbance at 570 nm was recorded using a spectrophotometer (Techcomp UV2300, Shanghai, China). The standard curve was conducted by using 50 μg/mL leucine solutions in parallel. Lysine content of each sample was calculated using the equation as follows: Lys content (g/100 g dry seed weight) = (measured lysine content × hydrolysis volume)/sample weight.

CONCLUSIONS

In summary, we obtained marker-free transgenic maize inbred X178 lines harboring a potato lysine-rich protein gene, *SBgLR*, and a tomato ERF transcription factor gene, *TSRF1*, via particle bombardment mediated co-transformation. Molecular analysis revealed that both of these two target genes were successfully expressed and showed various expression levels in different transgenic lines. Quantification of protein and lysine content in T_1 maize seeds showed that transgenic lines had both increased protein and lysine content. The protein and lysine content increased by 7.7% to 24.38% and 8.70% to 30.43%, compared to non-transformed maize, respectively. Moreover, transgenic maize exhibited more tolerance to salt stress. This study provides an effective way of corn molecular breeding and materials for improved nutritive quality and salt tolerance.

ACKNOWLEDGMENTS

We are grateful to Xiaojuan Wang and Hanwen Liang for assistance with the embryogenic calli preparation. This work was sponsored by the National Transgenic Major Program of China (Grant No. 2011ZX003-002) and the National Basic Research Program of China (2012CB215301).

CONFLICT OF INTEREST

Conflict of Interest

The authors declare no conflict of interest.

REFERENCES

1. Osborne T.B., Mendel L.B. Nutritional properties of proteins of the maize kernel. J. Biol. Chem.1914;18:1–16.
2. Mertz E.T., Bates L.S., Nelson O.E. A mutant gene that changes the protein composition and increases the lysine content of maize endosperm. Science. 1964;145:279–280.
3. Vasal S.K., Villegas E., Bjarnason M., Gelaw B., Goertz P. Genetic Modifiers and Breeding Strategies in Developing Hard Endosperm opaque-2 Materials. In: Pollmer W.G., Phipps R.H., editors. Improvement of Quality Traits of Maize for Grain and Silage Use. 2nd ed. Martinus Nijhoff; London, UK: 1980. pp. 37–73.
4. Segal G., Song R., Messing J. A new opaque variant of maize by a single dominant RNA interference-inducing transgene. Genetics. 2003;165:387–397.
5. Huang S., Adams W.R., Zhou Q., Malloy K.P., Voyles D.A., Anthony J., Kriz A.L., Luethy M.H. Improving nutritional quality of maize proteins by expressing sense and antisense zein genes. J. Agric. Food Chem. 2004;52:1958–1964.
6. Huang S., Frizzi A., Florida C.A., Kruger D.E., Luethy M.H. High lysine and high tryptophan transgenic maize resulting from the reduction of both 19- and 22-kD-zeins. Plant Mol. Biol. 2006;61:525–535.
7. Huang S., Kruger D.E., Frizzi A., D'Ordine R.L., Florida C.A., Adams W.R., Brown W.E., Luethy M.H. High-lysine corn produced by the combination of enhanced lysine biosynthesis and reduced zein accumulation. Plant Biotechnol. J. 2005;3:555–569.
8. Houmard N.M., Mainville J.L., Bonin C.P., Huang S., Luethy M.H., Malvar T.M. High lysine corn generated by endosperm specific

suppression of lysine catabolism using RNAi. Plant Biotechnol. J.2007;5:605–614.

9. Lucas D.M., Glenn K.C., Bu J.Y. Petition for determination of nonregulated status for lysine maize LY038. [(accessed on 13 August 2004)]. Available online:http://www.aphis.usda.gov/brs/aphisdocs/0422901p.pdf.
10. Yang S.H., Moran D.L., Jia H.W., Bicar E.H., Lee M., Scott M.P. Expression of a synthetic porcine-lactalbumin gene in the kernels of transgenic maize. Transgenic Res. 2002;11:11–20.
11. Bicar E.H., Woodman-Clikeman W., Sangtong V., Peterson J.M., Yang S.S., Lee M., Scott M.P. Transgenic maize endosperm containing a milk protein has improved amino acid balance. Transgenic Res.2008;17:59–71.
12. Liu J.Q., Seul U., Thompson R. Cloning and characterization of pollen-specific cDNA encoding a glutamic acid-rich protein (GARP) from potato *Solanum berthautii*. Plant Mol. Biol. 1997;33:291–300.
13. Yu J., Peng P., Zhang X., Zhao Q., Zhu D., Sun X., Liu J., Ao G. Seed-specific expression of a lysine rich protein sb401 gene significantly increases both lysine and total protein content in maize seeds. Mol. Breed. 2004;14:1–7. [PubMed]
14. Lang Z., Zhao Q., Yu J., Zhu D., Ao G. Cloning of potato SBgLR gene and its intron splicing in transgenic maize. Plant Sci. 2004;166:1227–1233.
15. Xiong L., Schumaker K.S., Zhu J. Cell signaling during cold, drought and salt stress. Plant Cell.2002;14:S165–S183.
16. Xiong L., Zhu J.K. Regulation of abscisic acid biosynthesis. Plant Physiol. 2003;133:29–36.
17. Guo H., Ecker J.R. The ethylene signaling pathway: New insights. Curr. Opin. Plant Biol. 2004;7:40–49.
18. Stockinger E.J., Gilmour A.J., Thomashow M.F. *Arabidopsis thaliana* CBF1 encodes an AP2 domain-containing transcriptional activator that binds to the C-repeat/DRE, a cis-acting DNA regulatory element that stimulates transcription in response to low temperature and water deficit. Proc. Natl. Acad. Sci. USA.1997;94:1035–1040.

19. Chakravarthy S., Tuori R.P., D'Ascenzo M.D., Fobert P.R., Despres C., Martin G.B. The tomato transcription factor Pti4 regulates defense-related gene expression via GCC box and non-GCC box cis elements. Plant Cell. 2003;15:3033–3050.
20. Guo Z.J., Chen X.J., Wu X.L., Ling J.Q., Xu P. Overexpression of the AP2/EREBP transcription factor OPBP1 enhances disease resistance and salt tolerance in tobacco. Plant Mol. Biol. 2004;55:607–618.
21. Yamaguchi-Shinozaki K., Shinozaki K. Transcriptional regulatory networks in cellular responses and tolerance to dehydration and cold stresses. Annu. Rev. Plant Biol. 2006;57:781–803.
22. Mantiri F.R., Kurdyukov S., Lohar D.P., Sharopova N., Saeed N.A., Wang X.D., Vandenbosch K.A., Rose R.J. The transcription factor MtSERF1 of the ERF subfamily identiWed by transcriptional prowling is required for somatic embryogenesis induced by auxin plus cytokinin in *Medicago truncatula*. Plant Physiol.2008;146:1622–1636.
23. Ohme-Takagi M., Shinshi H. Ethylene-inducible DNA binding proteins that interact with an ethylene-responsive element. Plant Cell. 1995;7:173–182.
24. Fujimoto S.Y., Ohta M., Usui A., Shinshi H., Ohme-Takagi M. Arabidopsis ethylene-responsive element binding factors act as transcriptional activators or repressors of GCC box-mediated gene expression. Plant Cell. 2000;12:393–404.
25. Gu Y.Q., Wildermuth M.C., Chakravarthy S., Loh Y.T., Yang C., He X., Han Y., Martin G.B. Tomato transcription factors pti4, pti5, and pti6 activate defense responses when expressed in Arabidopsis. Plant Cell.2002;14:817–831.
26. Liu Q., Kasuga M., Sakuma Y., Abe H., Miura S., Yamaguchi-Shinozaki K., Shinozaki K. Two transcription factors, DREB1 and DREB2, with an EREBP/AP2 DNA binding domain separate two cellular signal transduction pathways in drought- and low-temperature-responsive gene expression, respectively, in Arabidopsis. Plant Cell. 1998;10:1391–1406.
27. Hong J.P., Kim W.T. Isolation and functional characterization of the Ca-DREBLP1 gene encoding a dehydration-responsive element binding-factor-like protein 1 in hot pepper (*Capsicum annuum* L. cv. Pukang) Planta. 2005;220:875–888.

28. Liu N., Zhong N.Q., Wang G.L., Li L.J., Liu X.L., He Y.K., Xia G.X. Cloning and functional characterization of PpDBF1 gene encoding a DRE-binding transcription factor from *Physcomitrella patens*.Planta. 2007;226:827–838.
29. Zhang H., Zhang D., Chen J., Yang Y., Huang Z., Huang D., Wang X., Huang R. Tomato stress-responsive factor TSRF1 interacts with ethylene responsive element GCC box and regulates pathogen resistance to *Ralstonia solanacearum*. Plant Mol. Biol. 2004;55:825–834.
30. Zhang H., Li W., Chen J., Yang Y., Zhang Z., Wang X.C., Huang R. Transcriptional activator TSRF1 reversely regulates pathogen resistance and osmotic stress tolerance in tobacco. Plant Mol. Biol. 2007;63:63–71.
31. Zhou J., Zhang H., Yang Y., Zhang Z., Zhang H., Hu X., Chen J., Wang X.C., Huang R. Abscisic acid regulates TSRF1-mediated resistance to *Ralstonia solanacearum* by modifying the expression of GCC box-containing genes in tobacco. J. Exp. Bot. 2008;59:645–652.
32. Quan R., Hu S., Zhang Z., Zhang H., Zhang Z., Huang R. Overexpression of an ERF transcription factor TSRF1 improves rice drought tolerance. Plant Biotech. J. 2010;8:476–488.
33. Bhatnagar-Mathur P., Vadez V., Sharma K. Transgenic approaches for abiotic stress tolerance in plants: Retrospect and prospects. Plant Cell Rep. 2008;27:411–424.
34. James C. Global Status of Commercialized Biotech/GM Crops: 2010. ISAAA Brief No. 42. ISAAA; Ithaca, NY, USA: 2010.
35. Tuteja N., Verma S., Sahoo R.K., Raveendar S., Reddy B.L. Recent advances in development of marker-free transgenic plants: Regulation and biosafety concern. J. Biosci. 2012;37:167–197.
36. Holger P. Removing selectable marker genes: Taking shortcut. Trends Plant Sci. 2000;7:273–274.
37. De Neve M., de Buck S., Jacobs A., van Montagu M., Depicker A. T-DNA integration patterns in co-transformed plant cells suggest that T-DNA repeats originate from co-integration of separate T-DNAs. Plant J. 1997;11:15–29.
38. Sripriya R., Raghupathy V., Veluthambi K. Generation of selectable marker-free sheath blight resistant transgenic rice plants by efficient

co-transformation of a cointegrate vector T-DNA and a binary vector T-DNA in one *Agrobacterium tumefaciens* strain. Plant Cell Rep. 2008;27:1635–1644.

39. Miller M., Tagliani L., Wang N., Berka B., Bidney D. High efficiency transgene segregation in co-transformed maize plants using an *Agrobacterium tumefaciens* 2 T-DNA binary system. Transgenic Res.2002;11:381–396.

40. Shiva P.N., Bhojaraja R., Shivbachan S.K., Hari P.G.G., Nagraj T.K., Prasad V., Srikanth B.V., Jayaprakash T.L., Dasgupta S., Spencer T.M., et al. Marker-free transgenic corn plant production through co-bombardment. Plant Cell Rep. 2009;28:1655–1668.

41. Kumar S., Arul L., Talwar D. Generation of marker-free Bt transgenic indica rice and evaluation of its yellow stem borer resistance. J. Appl. Genet. 2010;51:243–257.

42. Qi Y., Ye S., Lu Y., Jin Q., Zhang X. Development of marker-free transgenic Cry1Ab rice with Lepidopteran pest resistance by *Agrobacterium* mixture-mediated co-transformation. Rice Sci. 2009;16:181–186.

43. Ramana R.M.V., Parameswari C., Sripriya R., Veluthambi K. Transgene stacking and marker elimination in transgenic rice by sequential *Agrobacterium*-mediated co-transformation with the same selectable marker gene. Plant Cell Rep. 2011;30:1241–1252.

44. Wang D., Zhao Q., Zhu D., Ao G., Yu J. Particle-bombardment-mediated co-transformation of maize with a lysine rich protein gene (sb401) from potato. Euphytica. 2006;150:75–85.

45. Schnable P.S., Ware D., Fulton R.S., Stein J.C., Wei F., Pasternak S., Liang C., Zhang J., Fulton L., Graves T.A., et al. The B73 maize genome: Complexity, diversity, and dynamics. Science. 2009;20:1112–1115.

46. Altschul S.F., Madden T.L., Schaffer A.A., Zhang J., Zhang Z., Miller W., Lipman D.J. Gapped BLAST and PSI-BLAST: A new generation of protein database search programs. Nucleic Acids Res.1997;25:3389–3402.

47. Nakano T., Suzuki K., Fujimura T., Shinshi H. Genome-wide analysis of the ERF gene family in*Arabidopsis* and rice. Plant Physiol. 2006;140:411–432.

48. Clustal W. [(accessed on 10 September 2007)]. Available online:http://www.ebi.ac.uk/Tools/clustalw2/index.html.
49. Song R., Segal G., Messing J. Expression of the sorghum 10-member kafirin gene cluster in maize endosperm. Nucleic Acids Res. 2004;32:e189.
50. Huang S., Jin L., Du J., Li H., Zhao Q., Ou G., Ao G., Yuan M. SB401, a pollen-specific protein from*Solanum berthaultii*, binds to and bundles microtubules and F-actin. Plant J. 2007;51:406–418.
51. Noble M., Lewis S., Cowan N.J. The microtubule binding domain of microtubule-associated protein MAP1B contains a repeated sequence motif unrelated to that of MAP2 and Tau. J. Cell Biol.1989;109:3367–3376.
52. Wiche G., Oberkanins C., Himmler A. Molecular structure and function of microtubule-associated proteins. Int. Rev. Cytol. 1991;124:217–273. [PubMed]
53. Kishor P.B.K., Sangam S., Amrutha R.N., Laxmi P.S., Naidu K.R., Rao K.R.S.S., Rao S., Reddy K.J., Theriappan P. Regulation of proline biosynthesis, degradation, uptake and transport in higher plants: Its implications in plant growth and abiotic stress tolerance. Curr. Sci. 2005;88:424–438.
54. Stepien P., Klobus G. Water relations and photosynthesi in *Cucumis sativus* L. leaves under salt stress.Biologia Plantarum. 2006;50:610–616.
55. Zhang H., Yang Y., Zhang Z., Chen J., Wang X.C., Huang R. Expression of the ethylene response factor gene TSRF1 enhances abscisic acid responses during seedling development in tobacco. Planta.2008;228:777–787.
56. Savoure A., Jaoua S., Hua X.J., Ardiles W., van Montagu M., Verbruggen N. Isolation, characterization, and chromosomal location of a gene encoding the DELTA-1-pyrroline-5- carboxylate synthetase in*Arabidopsis thaliana*. FEBS Lett. 1995;372:13–19.
57. Yoshiba Y., Kiyosue T., Katagiri T., Ueda H., Mizoguchi T., Yamaguchi-Shinozaki K., Wada K., Harada Y., Shinozaki K. Correlation between the induction of a gene for delta 1-pyrroline-5-carboxylate synthase and the accumulation of proline in *Arabidopsis thaliana* under osmotic stress. Plant J. 1995;7:751–760.

58. Zhang C.S., Lu Q., Verma D.P.S. Removal of feedback inhibition of delta 1-pyrroline-5- carboxylate synthase, a bifunctional enzyme catalyzing the first two steps of proline biosynthesis in plants. J. Biol. Chem.1995;270:20491–20496.

59. Strizhov N., Abraham E., Okresz L., Blickling S., Zilberstein A., Schell J., Koncz C., Szabados L. Differential expression of two P5CS genes controlling proline accumulation during salt-stress requires ABA and is regulated by ABA1, ABI1 and AXR2 in Arabidopsis. Plant J. 1997;12:557–569.

60. Sambrook J., Fritsch E.F., Maniatis T. Molecular Cloning: A Laboratory Manual. 2nd ed. Cold Spring Harbor Press; New York, NY, USA: 1989.

61. Saghai-Maroof M.A., Soliman K.M., Jorgensen R.A., Allard R.W. Ribosomal DNA spacer-length polymorphisms in barley: Mendelian inheritance, chromosomal location, and population dynamics. Proc. Natl. Acad. Sci. USA. 1984;81:8014–8018.

62. Bates L.S., Waldren R.P., Teare I.D. Rapid determination of free proline for water-stress studies. Plant Soil. 1973;39:205–207.

63. Walker N.J. A technique whose time has come. Science. 2002;296:557. [PubMed]

64. geNorm v3.5 software. [(accessed on 7 May 2002)]. Available online:http://medgen.ugent.be/~jvdesomp/genorm/

Chapter 7

Application of Genomic Tools in Plant Breeding

A.M. Pérez-de-Castro, S. Vilanova, J. Cañizares, L. Pascual, J.M. Blanca, M.J. Díez, J. Prohens, and B. Picó

Instituto de Conservación y Mejora de la Agrodiversidad Valenciana, Universitat Politècnica de València, Camino de Vera 14, 46022 Valencia, Spain.

ABSTRACT

Plant breeding has been very successful in developing improved varieties using conventional tools and methodologies. Nowadays, the availability of genomic tools and resources is leading to a new revolution of plant breeding, as they facilitate the study of the genotype and its relationship with the phenotype, in particular for complex traits. Next Generation Sequencing (NGS) technologies are allowing the mass sequencing of genomes and transcriptomes, which is producing a vast array of genomic information. The analysis of NGS data by means of bioinformatics developments allows discovering new genes and regulatory sequences and their positions, and makes available large collections of molecular markers. Genome-wide expression studies provide breeders with an understanding of the molecular basis of complex traits. Genomic approaches include TILLING and EcoTILLING, which make possible to screen mutant and germplasm collections for allelic variants in target genes. Re-sequencing of genomes is very useful for the genome-wide discovery of markers amenable for high-throughput genotyping platforms, like SSRs and SNPs, or the construction of high density genetic maps. All these tools and resources facilitate studying the genetic diversity, which is important for germplasm management, enhancement and use. Also, they allow the identification of markers linked to genes and QTLs, using a diversity of techniques like bulked segregant analysis (BSA), fine genetic mapping, or association mapping. These new markers are used for marker assisted selection, including marker assisted backcross selection, 'breeding by design', or new strategies, like genomic selection. In conclusion, advances in genomics are providing breeders with new tools and methodologies that allow a great leap forward in plant breeding, including the 'superdomestication' of crops and the genetic dissection and breeding for complex traits.

INTRODUCTION

Ever since the beginnings of the domestication of plants, some 10,000 years ago, plant breeding has been extremely successful in developing crops and varieties that have contributed to the development of modern societies, and have successively beaten (neo-)Malthusian predictions [1]. Application of conventional pre-genomics scientific breeding methodologies has led to the development of modern cultivars, which have contributed to the dramatic improvement of yield of most major crops since the middle of the 20th century. The success of plant breeding in the last century has relied in the utilization of natural and mutant induced genetic variation and in the efficient selection, by using suitable breeding methods, of the favorable genetic combinations. In this respect, the evaluation and identification of genetic variants of interest as well as the selection methodologies used have largely been based in the phenotypic evaluation.

Nowadays, genomics provides breeders with a new set of tools and techniques that allow the study of the whole genome, and which represents a paradigm shift, by facilitating the direct study of the genotype and its relationship with the phenotype [2]. While classical genetics revolutionized plant breeding at the beginning of the 20th century, genomics is leading to a new revolution in plant breeding at the beginning of the 21thcentury.

The field of genomics and its application to plant breeding are developing very quickly. The combination of conventional breeding techniques with genomic tools and approaches is leading to a new genomics-based plant breeding. In this new plant breeding context, genomics will be essential to develop more efficient plant cultivars, which are necessary, according to FAO, for the new 'greener revolution' needed to feed the world's growing population while preserving natural resources.

One of the main pillars of genomic breeding is the development of high-throughput DNA sequencing technologies, collectively known as next generation sequencing (NGS) methods. These and other technical revolutions provide genome-wide molecular tools for breeders (large collections of markers, high-throughput genotyping strategies, high density genetic maps, new experimental populations, etc.) that can be incorporated into existing breeding methods [2-5]. Recent advances in genomics are producing new plant breeding methodologies, improving and accelerating the breeding process in many ways (e.g., association mapping, marker assisted selection, 'breeding by design', gene pyramiding, genomic selection, etc.) [5-7].

Genomics approaches are particularly useful when dealing with complex traits, as these traits usually have a multi-genic nature and an important environmental influence. Thanks to these technological improvements it is now feasible for a small laboratory to generate in a short time span (e.g., several months) enough molecular data to obtain a set of mapped quantitative trait loci (QTLs), even in a species lacking any previous genomic information [8]. Genomic tools are thus facilitating the detection of QTLs and the identification of existing favorable alleles of small effect, which have frequently remained unnoticed and have not been included in the gene pool used for breeding [9, 10]. In this review, we present and discuss the most relevant advances in the development and application of genomic tools for plant breeding, in particular for complex traits. Firstly, we introduce the most relevant genomic tools and secondly we provide examples of the application of these tools to plant breeding. The objective is to provide modern breeders with an updated synthetic view of how genomics can effectively improve the efficiency of breeding programs.

GENOMIC TOOLS AND RESOURCES FOR PLANT BREEDING

Genome and Transcriptome Sequencing

The availability of the whole genome sequence of a crop is of great utility for plant breeding, but until recently it has been an unachievable goal for most crops. This privilege was restricted to a reduced number of model species with small genomes and to projects with an important investment in time and resources, but now has extended to an increasing number of crops. However, it is also true that for important cultivated species with large and complex genomes such as wheat, sugarcane, or coffee, the sequencing of the whole genome is very challenging and may take several years before a draft is completed. Given the high cost of whole genome sequencing, transcriptome sequencing has been a cheaper alternative. The cDNA sequences (expressed sequence tags, ESTs) provide relevant information about the genes expressed in a certain tissue or organ, at a given stage of development and under particular environmental conditions. ESTs sequencing projects do not provide information about non-coding sequences and, even using diverse libraries, it is difficult to identify all genes and transcripts variants. Despite these limitations, ESTs collections have been very useful for breeders.

The Sanger technology has been the predominant sequencing method for the past thirty years. It has been used to sequence several genomes as well as many transcriptomes. The first international collaborative project resulted in the whole genome sequence of the model plant *Arabidopsis thaliana* [11]. Since then, reference genomes of selected genotypes were completed in a limited number of crops such as rice [12], maize [13], sorghum [14], populous [15], grapevine

[16], papaya [17], or soybean [18]. The transcriptomes of most major crops, to a greater or lesser extent, were also sequenced. A global view of the genomes and transcriptomes sequenced can be obtained from the Gene Index Project (http://compbio.dfci.harvard.edu/tgi/plant.html) or in the NCBI Unigene database (http://www.ncbi.nlm.nih.gov/unigene).

The field of genomics has changed with the arrival of NGS technologies [19]. These new technologies have reduced the cost of sequencing by more than one thousand times compared to Sanger technology, by avoiding time-consuming and tedious traditional cloning steps and making possible to perform millions of sequencing reactions in parallel (Table1). Among the "second generation" technologies, the 454 (Roche,http://www.454.com) and Illumina (Illumina, http://www.illumina.com) platforms are already profusely used to sequence crop species. Others, like Solid (Applied Biosystems,http://www.appliedbiosystems.com/technologies), have been less exploited in plants. By using these NGS technologies, the sequencing capacity of laboratories is continuously increasing. For instance, one High-Seq 2000 Illumina Sequencer is able to generate 55 Gb per day, which is roughly eighteen times the size of the human genome. Moreover, new, "third generation" platforms are being developed and incorporated to sequencing projects, such as PacBio RS (Pacific Biosciences, http://www.pacificbiosciences.com), Helicos (Helicos, http://www.helicosbio.com), or Ion Torrent (Life Technologies, http://www.iontorrent.com). The sequence obtained by NGS are generally deposited in the NCBI Sequence Read Archive (http://www.ncbi.nlm.nih.gov/unigene).

Table 1: Comparison of the Main Characteristics of the Conventional Sanger and Some of the Most Currently Used Next Generation Sequencing (NGS) Technologies and Approximate Sequencing Cost (in US $ Per Mbp)

Technology	Read Length (bp)	Mbp per run	Cost ($/Mbp)
Sanger	1000	0.001	3000.00
454 Roche	450	450	66.00
Illumina Hi-Seq2000	100	270000	0.07
Solid 5500xl	50	270000	0.07

Nowadays, it is feasible to sequence most crop genomes (excluding those with a very large and complex genome) with a relatively modest budget, by combining Sanger with NGS technologies. Sequencing projects for 135 plant genomes, most of them corresponding to cultivated species or wild relatives, have been completely sequenced (3), are being assembled (27) or are in progress (105), as reported at the NCBI database (http://www.ncbi.nlm.nih.gov/genomes/leuks.cgi). Other databases including plant genome sequences are Plant GDB (http://www.plantgdb.org) and Phytozome (http://www.phytozome.net). A fully sequenced and well annotated genome provides useful tools for the breeders, as it allows the discovery of genes, determining their position and function, as well as the development of large marker collections and high resolution maps. In the cases where only a draft sequence is available, its usefulness depends on the quality of the assembly. For instance on many occasions thousands of scaffolds are obtained but they are not anchored to the genetic map, which difficults the use of the information. Many transcriptomes have also been sequenced, a number of them in species for which no previous sequence information was available. Sweet potato [20], squash [21], pigeonpea [22], or buckwheat [23] represent some examples published in the last months. These assays are showing the great complexity of plant transcriptomes, allowing the identification of rare transcript variants that are being used to improve gene annotation and our knowledge of gene function and regulation.

Bioinformatics

NGS technologies are facilitating sequencing projects, but have brought new challenges, as millions of short DNA reads have to be analysed and assembled [19]. Also, genetic maps, genotypes, or expression information at a genomic scale have to be processed in order to obtain the relevant biological information. Therefore, it is necessary to develop new bioinformatics tools (algorithms and software), which allow the analyses of huge amounts of genome-wide data, but it is also necessary to change the approaches used to understand complex biological traits [25, 26].

The field of sequence analysis has a long tradition and has enabled the assembly of many genome sequences obtained by Sanger sequencing. Nowadays, the huge amount of sequence reads generated by NGS and the low quality of individual reads requires new software tools and algorithms that allow dealing with these data in an efficient way [27]. We consider this to be a limiting factor for this kind of analyses right now. Although in the last years great advancements in the sequence processing algorithms have been achieved, the software currently available requires improvements in many cases.

Two of the most common analyses carried out on these NGS reads are genome assembly and annotation and mapping. Genome assembly is a complex task requiring powerful computers and skilled bioinformaticians [25]. In particular, the RAM memory requirements of the computers used to assemble eukariotic genomes could hinder the application of this technique by small laboratories. Some of the most commonly used assemblers are Roche›s 454 Gsassembler, Celera Assembler, and Mira. Once a reference genome is available in the species it is common to study its variation [19]. To do this, mapper software is commonly used instead of an assembler software. A mapper tries to align every read against the reference genome. This process is much simpler and faster than the assembly. In this case the computer requirements are usually less demanding and the limiting factor could be the storage capacity. Some commonly used mappers are Bowtie, BWA, and TopHat. Once the reads are aligned, single nucleotide polymorphism (SNPs) can be detected by using the SAMtools or the GigaBayes SNP callers [28].

The open source software mainly used by the bioinformaticians is cumbersome for users not well versed in the Unix command line operating system. Some commercial proprietary solutions easier to use have been developed (e.g., LaserGene or CLC Genomics Workbench), but they have not been widely embraced by the breeders. Galaxy is an open source project devoted to create an easy to use web interface to the open source CLI based applications used in this area.

An important amount of work has been devoted in this field to the creation of standard and open file formats capable of storing information regarding sequence alignment and modelling (SAM) [24], SNP calls using variant call format (VCF; http://1000genomes.org/wiki/doku.php?id=1000_genomes: analysis:vcf4.0), genomic regions with browser extensible data (BED, http://genome.ucsc.edu/FAQ/FAQformat# format1) and genomic annotations using the general feature format (GFF;http://www.sequenceontology.org/resources/gff3.html]). These open and standard formats allow the interoperability of the different software tools that are being actively developed and used. In addition, the computer requirements might be strong as some analyses require a large amount of RAM memory or storage capability.

The algorithms and methods used to store and process raw genomic data generated by the different technological platforms will depend on the type of data being processed and on the result expected. In any case, once the relevant information is obtained by the bioinformaticians, results have to be made available to the breeders by using an interface as easy and friendly as possible [25]. To provide access to this information, the generation of an easily browseable web site is a common and usually successful approach. Several general purpose web databases exist to make the relevant biological information available to the researchers and breeders (Table 2), like GenBank (http://www.ncbi.nlm.nih.gov/genbank/), EBML (http://www.ebi.ac.uk/embl/), DDBJ (http://www.ddbj.nig.ac.jp/) and Swiss-prot (http://expasy.org/sprot/). These latter databases are devoted to store information about any species, but several other more specific databases focused on species of interest to the breeders also exist, like the Sgn (http://solgenomics.net/), Phytozome (http://www.phytozome.net/), Gramene (http://www.gramene.org/) or CropNet (http://ukcrop.net/), which hold information that could have more specific use for breeding programs.

Table 2: Some Important Databases and Repositories of Genomic Information of Interest for Breeders

Technology	Read Length (bp)	**Mbp per run**	**Cost ($/Mbp)**
Sanger	1000	0.001	3000.00
454 Roche	450	450	66.00
Illumina Hi-Seq2000	100	270000	0.07
Solid 5500xl	50	270000	0.07

Expression Studies, from Microarrays to RNA-seq

New genomic tools are also of interest to expand and accelerate gene expression studies. The analysis of gene expression has provided a rich source of biological information, which allows breeders to understand the molecular basis of complex plant processes, leading to the identification of new targets for manipulating these processes.

Gene expression studies were at first based on the classical Northern blot method that only allowed the quantification of tens of genes simultaneously. The QRT-PCR is a more affordable and quantitative technique; but the number of genes analyzed by experiment is also limited [29]. Other approaches allowing the study of thousands of genes were differential display and cDNA amplified fragment length polymorphisms (cDNA-AFLPs) [30]. However, these methods are not really quantitative and are limited by the ability of the developed libraries to capture low-abundance transcripts. Other methods that overcome part of these problems are the serial analysis of gene expression (SAGE) [31] and massively parallel signature sequencing (MPSS) [32]. Nevertheless, the most employed methods at present to analyze transcript profiling are the hybridization-based platforms or microarrays [33]. Expression arrays have several advantages when compared with other methods. They can measure tens of thousands of different transcripts in the same reaction, they are semi-quantitative and sensitive to low-abundance transcripts if those are represented in a given array.

There are several web resources that facilitate microarray data analysis (e.g., http://babelomics.bioinfo.cipf.es/) [34] or even web pages where the breeder can download experiments already performed and analyzed. There are also software packages specializated in microarrays analysis as the Bioconductor (http://www.bioconductor.org/help/work-flows/oligo-arrays/) or MeV (ttp://www.tm4.org/mev/) [35]. Probably one of the most useful database is Genevestigator (https://www.genevestigator.com/gv/ doc/plant_biotech.jsp) [36], which contains microarray data from different species. The most extensive data are from the model species *A. thaliana* [37], but an increasing number of studies in crops like maize, wheat, rice, barley, or soybean are already available. All published expression data are public and disposables in databases as GEO (http://www.ncbi.nlm.nih.gov/geo/) [38], Array Express (http://www.ebi.ac.uk/array-express/) [39] or species specific databases. These data can be really useful when analyzing gene expression in these species or other crops [40].

Microarrays make use of the existing EST collections and genome sequence data. The vast increase provided by NGS in the number of sequences opens the possibilities of expression studies in a large number of species lacking previous sequence information. Also, deep NGS sequencing of RNA transcripts (RNA-seq) is emerging as an alternative to microarray studies to quantify gene expression [41, 42]. RNA-seq does not depend on genome annotation or on the probes contained in the array platform. This technology is also very useful to improve genome annotation, improving the detection of rare transcripts and splicing variants and the mapping of exon/intron boundaries. Moreover, RNA-seq avoids bias introduced during hybridization of microarrays and saturation level problems, haa a greater sensibility, and shows high reproducibility [41, 43]. This approach has been already used in different crops with different breeding objectives, leading to the identification of genes involved in several metabolic pathways, disease response, fruit development, etc. [44-47]. All these studies show the potential of RNA-seq for complex traits breeding. However, RNA-seq is an emerging technology and the methods used to analyze this kind of data are still being developed.

Mutant and Germplasm Collections in the Genomics Era: TILLING and EcoTILLING

Plant breeding requires genetic variability to be selected in order to increase the frequencies of favourable alleles and genetic combinations. Sources of natural genetic variability can be found within the crop, mostly in the form of landraces, and also in the wild relatives. Although many landraces have been substituted by modern and uniform cultivars and genetic erosion has taken place in wild materials, gene banks preserve many of these materials, which constitute an important reservoir of genetic variation useful for breeding [48].

Another important source of variability corresponds to the artificial mutant collections developed in several crop species. These artificial collections are created by radiation, chemical mutagenesis, or transgenic and insertion methods. Artificial mutations, mostly obtained by radiation and chemical methods, were used in plant breeding in the pre-genomics era, but new technologies are allowing the development of other types of collections [49]. For instance, the transferred DNA tagged lines and transposon tagged lines have been used to develop mutant collections in several species such as the model plant *Arabidopsis* (The

Arabidopsis Information Resource; http://www.arabidopsis.org) or rice (International Rice Functional Consortium;http://irfgc.irri.org). Gene silencing technologies, using RNA interference, have also been used to create gene specific mutant collections in *Arabidopsis*, like the AGRIKOLA project (http://www.agrikola.org). The artificial mutant collections frequently contain variability not present in the natural collections, and so are also very useful for the study and development of new traits.

In order to facilitate the identification of the accessions of interest in these collections, a genetic reverse approach has been used. Targeting Induced Local Lesions in Genomes (TILLING) [50] is able to identify all allelic variants of a DNA region present in an artificial mutant collection. A similar procedure called ecotype TILLING (EcoTILLING) [51] can be used to identify allelic variants for targeting genes in natural collections. These two methods are based on the use of endonucleases, such as CEL I or Endo I, that recognize and cut mismatches in the double helix of DNA [52, 53]. Since the TILLING and EcoTILLING techniques identify all allelic variants for a certain genomic region, the phenotypic characterization effort can be concentrated in a reduced number of accessions with different variants. Obviously, the success of the identification of variation useful for breeding programmes will depend on the right selection of target genes. The availability of sequences coming from NGS sequencing projects and the information provided by gene expression studies is significantly increasing the number and quality of candidates for TILLING and EcoTILLING studies.

These procedures have been successfully used in many crops [54]. For example, TILLING has been applied to *Arabidopsis* [55], *Lotus* [56], barley [57], maize [58], pea [59], and melon [60]. Rice was the first crop for which EcoTILLING was applied [61]. Subsequently, EcoTILLING has been used in other crops and wild relatives, like barley [62], wheat [63], or the wild peanut *Arachis duranensis* [64], using both genebank collections and natural populations [65]. These assays used gene targets involved in different processes. Many studies have been focused on detecting allelic variants in genes most related to organoleptic quality [66, 67] or disease resistance [68, 69].

Re-Sequencing for SNPs Discovery and Use in Genotyping Platforms

One of the most interesting applications of NGS for plant breeders is the discovery of genetic variation. Now it is possible to sequence rapidly multiple individuals within a species with limited technical expertise and at minimal cost. The parallel development of computational pipeline tools is greatly accelerating the accurate mining of these sequences for genetic variants that can be converted into genetic markers, mainly microsatellites or simple sequence repeats (SSRs) and SNPs [70]. SSRs and SNPs are now the predominant markers in plant genetic analysis. SNPs are more abundant, stable, amenable to automation, and increasingly cost-effective, thus are fast becoming the marker system of choice in modern genomics research [71].

The genome-wide SNPs discovery by massive re-sequencing has been performed in model species with small genomes, such as *Arabidopsis thaliana*, where the 1001 Genomes project (http://www.1001genomes.org) [72] is attempting to unveil the whole-genome sequence variation in this reference plant. Similar re-sequencing efforts are becoming possible in rice, maize, grape, soybean, poplar etc. by sequencing sets of related genotypes, individually or pooled, within each species (elite cultivars, breeding lines, ecotypes, landraces, and/or weedy and wild relatives of a crop) [73-76]. The higher complexity in architecture and repeat content of these genomes has made necessary the use of approaches for genomic complexity reduction that also reduce sequencing cost [77]. Identification of SNPs is also very challenging in species with high levels of heterozygosity and/or with complex ploidy levels, as pseudo-SNPs are identified by misassembly of paralogs [78-80].

Both Roche 454 and Illumina GA have been mostly used for genome re-sequencing. The aligment difficulties often associtated to the use of short Illumina GA reads (Table **1**) are less problematic in species for which available reference genomes facilitates SNPs calling and genome positioning of genetic variation [81]. For most of these species, limited collections of SSRs and SNPs were available from early re-sequencing efforts, previous to the advent of NGS, but new genome-wide re-sequencing is enlarging the SNP pools and making them more representative of the range of natural variation.

For an increasing number of species with high societal or economic value NGS genome re-sequencing is providing the first SSRs and SNPs resources. Examples are the grain amaranths (Amaranthus sp.), important pseudocereals, appreciated for their nutritional quality and tolerance to abiotic stress [82], for which no prior genome information existed. In these cases the combination of several sequencing techniques, and the use of paired-end sequencing facilitates SNP discovery. Roche 454 and Illumina GA were combined for high-throughput SNP discovery in common bean [83] and also Solid was used to sequence diploid wheat species, which are donors of the subgenomes of modern hexaploid bread wheat [84].

Most of the effort in species lacking genomic resources is being made through transcriptome re-sequencing, as an alternative way of genome complexity reduction. Targeted amplicon re-sequencing is another strategy for discovering SNPs in specific genes [78].

One of the first examples of deep transcriptome sequencing was a study with two maize inbred lines [85]. This first study was followed by a large and rapidly increasing number of projects using non-model crops, some of them with large, complex, polyploid, and uncharacterized genomes, including forest trees, like *Eucalyptus* [86], oak [87], several polyploid crops, like oilseed rape [79], oats [80], coffee [88], and sweet potato [89], non-model grain legumes as chickpea and chickling pea [90], tomato [91], or several cucurbits, including *Cucurbita* spp. [21], cucumber [92], and melon [93].

These studies employ normalized/non-normalized cDNA libraries generated from single or multi-tissues samples, and derived from single or pooled genotypes, combined or not with multiplex identifier barcodes that allow allele origin identification. Sequencing is often focused on selected genotypes subjected to specific conditions, to detect SNPs in candidate genes involved in complex biological processes of interest to plant breeders. For example, the transcriptomes of two resistant and one susceptible lines of water yam, a major staple crop in Africa, under anthracnose infection, were successfully sequenced detecting SNPs in genes putatively involved in pathogen response [94]. Also, two alfalfa genotypes contrasting for cellulose and lignin content were sequenced, which allowed selecting SNPs useful to improve alfalfa as a forage crop and cellulosic feedstock [95]. The use of genome and transcriptome sequencing for SNP discovery has resulted in large SNPs collections in most of the crops. These large collections are being validated and applied for different purposes such as map construction, map saturation, genome-wide diversity studies, association mapping etc. (Table 3). Some of the most important achievements will be described in later sections.

Table 3: Some Examples of the Utility of Molecular Markers Developed by Means of High-throughput Genomics Techniques for the Breeding of Important Crops

Crop	Markers	**Plant Material**	**Use for Breeding**	**Reference**
Rice (*Oryza sativa*)	~$3.6 \cdot 10^6$ SNPs	517 rice landraces	Association studies for 14 agronomic traits	[96]
Barley (*Hordeum vulgare*)	1,536 SNPs	768 breeding lines	Association studies for *Fusarium* head blight resistance	[97]
	3,072 SNPs	336 DH lines and 213 germplasm selections	High-density genetic map construction and MAF estimation	[98]
Maize (*Zea mays*)	8,590 SNPs	553 elite maize inbred lines	Association studies for oleic acid content	[99]
	1,106 SNPs	5,000 RILs	Association studies for resistance to southern leaf blight	[100]
	1,536 SNPs	154 maize inbred lines	Diversity studies	[101]
Grapevine (*Vitis vinifera*)	94 SNPs and 7 indels	148 grape varieties	Association studies for muscat flavor candidate gene VvDXS	[102]
	9000 SNPs	10 cultivated *Vitis vinifera* and 7 wild *Vitis* spp.	Diversity and population structure studies	[74]
Pea (Pisum sativum)	384 SNPs	91 RIL mapping population and 373 Pisum accessions	Linkage map construction and diversity studies.	[103]

Wheat (*Triticum aestivum*)	874 DArT markers	winter wheat core collection of 96 accessions	Association studies for 20 agronomic traits	[104]
	1,536 SNPs	478 spring and winter wheat cultivars	Diversity studies	[105]
White spruce (*Picea glauca*)	944 SNPs	492 individuals	Association studies with 549candidate genes and 25 wood quality traits	[106]

There are many SNPs genotyping techniques, which are more or less appropriate for different scales of individuals/SNPs to be genotyped [107]. The implementation of marker-assisted breeding strategies often requires the generation of thousands of genotypes per population. Thus, one practical way of optimizing the use of these large SNPs collections is using them with cost-effective platforms for medium to high density genotyping. A large number of commercial platforms are available for semi automated or fully automated SNP genotyping [108, 109]. Genotyping assays usually require a previous process of selection of a set of SNPs, among the hundreds/tens of thousands detected, that are appropriate for the assay objectives.

The Illumina GoldenGate assays have been the most widely used for mid-throughput applications. SNPs platforms with 384, 768, or 1536 SNPs are available for a number of species (Table 3). Popularity is also increasing for the Sequenom Mass array and the KASPar genotyping chemistry [82, 110]. Expanded arrays with tens of thousands SNPs for high-throughput applications have been also developed with the Infinium technology in maize, grape, tomato, pine, and poplar and are under development in soybean and severalRosaceae crops (apple, peach, and cherry) [74, 111].

Construction of High Density Genetic Maps

One of the main applications of genomic advances is the development of high density genetic maps. The high-density map construction involves the location of hundreds or even thousand markers in the different linkage groups. In these maps the coverage should be very high and no large gaps must be present. NGS technologies and high-throughput genotyping platforms have allowed the improvement of genetic maps by increasing markers density. Several works include the integration of new markers, basically SNPs derived from re-sequencing studies, into previously developed genetic maps, both in diploid and polyploidy species [80, 112]. Golden Gate has been the most widely used platform. It has been estimated that this genotyping platform is 100-fold faster than gel-based methods for increasing 2-3 times maize map density [101]. Also Sequenom-based SNP-typing assays are starting to be applied. In a recent study, a total of 1.016 SNPs, identified via comparative next-generation transcriptomic sequencing, were successfully mapped by genotyping 297 maize recombinant inbred lines (RILs) [113]. Other genotyping strategies based on arrays hybridization, such as the single-feature polymorphisms (SFP), variants detected by a single probe in oligonucleotide arrays, are speeding up genetic map construction. This technique has been used for the construction of a high-density linkage map in species poorly characterized, like Eucalyptus [114]. The newly developed maps, enriched in sequence-based markers are facilitating comparative mapping. Recent examples are high density SNPs maps of barley compared with wheat and rice [98, 115].

The decrease of sequencing costs is also allowing the detection of new types of genetic markers useful for increasing the density of genetic maps.In this respect, restriction-site associated DNA (RAD) is a kind of marker which detects genetic variation adjacent to restriction enzyme cleavage sites across a target genome. These markers are produced after NGS sequencing of genomic libraries obtained after digestion with different restrictases. As an example of the utility of this technique, a total of 445 RAD markers distributed across all seven barley chromosomes were located, which was very useful for linkage map construction in this crop [116].

The markers derived from NGS can also be useful to position sequence scaffolds onto physical maps. In this respect, a new method combining deep sequencing and the bin mapping strategy has been described [117].The SNPs identified by re-sequencing genomic libraries from selected progeny individuals, derived from a cross between two closely related diploid strawberry species, were used to anchor 92.8% of the *Fragaria*genome to the genetic map. Results highlighted the potential of this methodology to obtain a robust framework for the anchoring of the genome sequence without the requirement of a high density physical mapping or a well-saturated genetic map. Whole-genome re-sequencing at different coverage levels is being increasingly applied for map construction using different strategies. As an example, a genetic map for rice has been constructed using whole genome re-sequencing of 150 RILs [118]. These authors concluded that the sequencing-based method was approximately 35 times more precise in recombination breakpoint determination than PCR-based markers maps. Also, the whole genome of 128 chromosome segment substitution lines (CSSLs) of rice was re-sequenced and used it for the construction of an ultra-high quality physical map in this crop [119]. Based on low coverage re-sequencing, a new mapping strategy that allows inferring the parental genotypes of the assayed RILs population has been proposed [120]. An ultra-high density linkage map was obtained with this method and the quality of the map was evaluated by using it to identify a QTL controlling grain width. Further applications of new sequence-based denser genetic maps to QTL discovery and marker assisted selection (MAS) will be discussed later.

TOWARDS A GENOMICS-BASED PLANT BREEDING

Genome-Wide Genetic Diversity Studies

One of the main challenges in agricultural genetics is to access and use the tremendous genetic variation present in germplasm collections and in the wild relatives. A significant part of this variation remains untapped because of the difficulties in effectively identifying genetic differences in large collections. Some traits, with high heritability and of simple characterization, are easy to select for. However, desirable allelic variants and genetic combinations for complex traits are difficult to identify. Recent advances in genotyping are enabling genome-wide diversity studies capable of better capturing the spectrum of variability in natural and breeding populations.

Most of the mid- to high-throughput genotyping platforms described above are being used for studies on diversity and population structure in the corresponding crops (Table 3). By using representative diversity panels, polymorphism rates for individual SNP markers, minor allele frequencies (MAFs), etc. are estimated, facilitating the selection of those SNPs with biological interest and highly polymorphic in the different groups. For example, the Infinium arrays developed in some of these crops are being used to create haplotype maps for vast germplasm collections, such as the 18,000 accessions of the USDA soybean germplasm collection [121].

Haplotype maps (hapmap) of entire collections are useful to identify rare, potentially valuable, alleles. Hapmap projects are undergoing in a number of species such as the "rice diversity project" (http://www.ricehapmap.org/index.aspx) aimed to develop a 10,000 SNP chip for rice and create a haplotype map to document the differences in allelic variation within and between the different subpopulations of O. sativa and its progenitor O. rufipogon. Large-scale genetic diversity studies have also been accomplished in maize. Gore et al. [122] identified and genotyped several million sequence polymorphisms among 27 diverse maize inbred lines. This study allowed the discovery of regions with highly suppressed recombination that appear to have influenced the effectiveness of selection during maize inbred development and may be a major component of heterosis. Also, highly differentiated regions were found that probably contained locithat are key to geographic adaptation. Also in legumes, the Medicago HapMap Project, that consist in the sequencing the whole-genomes of 30 inbred lines, will explore the genetic basis of symbiosis, creating a robust platform for genome-scale association mapping [123].

The diversity panels can include representatives of close or more distantly related species to check if these sequence-based SNP assays also work in related species [74, 82]. Sometimes sets of SNPs specifically developed for detecting genetic diversity among closely related cultivars are used in genotyping platforms. For example, despite the large amounts of SNPs available in rice obtained from the comparison of the two reference genomic sequences (one japonica and one indica variety) [124], extremely low levels of DNA polymorphism were detected within japonica cultivars. A whole-genome sequencing of an elite Japanese rice cultivar, closely related to the reference japonica variety, has been conducted and the SNP information obtained by comparison of the two japonica sequences was applied to develop a high-throughput genotyping array used for genotyping a set of representative Japanese cultivars [125]. These experiments are useful for understanding the role of selection and breeding in the distribution of genetic variation across the crop genome. In fact, this assay led to the identification of several haplotype blocks which are inherited from traditional landraces to current improved varieties. Moreover, it was found that, as predicted, modern breeding practices have generally decreased genetic diversity. On the practical level, the distribution of genetic diversity in modern cultivars plays an important role in the choice of specific mapping and crop improvement strategies.

Genome-wide survey of genetic diversity is useful to elucidate the causative genetic differences that give rise to observed phenotypic variation providing a foundation for dissecting complex traits through genome-wide association studies. However, its utility is limited if phenotypic data are not available. Not just genomics and transcriptomics, but the other ‹omics› disciplines, like proteomics and metabolomics, are useful to understand how the changes in the genotype lead to differences in the final phenotype. Phenomics, which uses high-throughput technologies to characterize germplasm, is being developed and will help to deal with this issue [126].

Identification of Molecular Markers Linked to Single Genes and QTLs

NGS and high-resolution maps have led to a significant improvement in the identification of molecular markers linked to specific genes and to QTLs. The most important advantage comes from the dense genome coverage, which allows the identification of markers closely linked to any target genomic region, with the advantages that this tight linkage provides.

Methods already used in the pre-genomics era to facilitate the identification of markers linked to single loci, such as bulked segregant analysis (BSA), are now optimized. For example, a GoldenGate assay has been combined with BSA to significantly accelerate mapping of the dominant resistance locus to soybean rustRpp3 [127]. In this respect, there is an increasing number of reports on exploitation of NGS technologies to identify molecular markers tightly linked to major genes. For example, a fine genetic mapping of the single dominant scab resistance gene (Ccu) in RILs of cucumber (Cucumis sativus) has been conducted [128]. The resistant cucumber genome was sequenced with

Solexa/Illumina NGS and compared with the susceptible cucumber genome. As a result, three insertion/deletion (indel) markers closely linked to the Ccu locus where obtained. A detailed study of the region delimited by markers revealed four resistance gene analogs as possible candidates for Ccu.

QTL detection has traditionally been conducted by linkage mapping. NGS technologies are significantly contributing to increase accuracy in detection of QTLs. They allow increases in many orders of magnitude of the number of markers mapped, ensuring high mapping resolution, and also aid in the development of mapping populations, such as RILs, near isogenic lines (NILs), and CSSLs, more appropriated for QTLs detection. These populations have conventionally been constructed and genotyped using a limited number of markers.

There are increasing reports describing accurate QTLs mapping with different NGS or high-throughput genotyping strategies. For example, a high density rice map constructed by whole-genome re-sequencing of a RILs population, was used to identify four QTLs controlling plant height [90]. On a different study [129] an ultra-high density genetic map based on SNPs, obtained with Illumina GA, was compared with a linkage map based on RFLPs/SSRs in rice. The positions of several cloned genes, two major QTLs for grain length and grain width, and a QTL for pigmentation were evaluated in a RIL population, arising the expected result that the SNPs map detected more QTLs and more accurately than a RFLPs/SSRs based linkage map.

QTL detection based on the linkage analysis method has the disadvantage that the number of recombination events is limited to the generations needed to develop the mapping population. Association mapping or linkage disequilibrium (LD) mapping is a new powerful approach to map complex traits. This method identifies genetic loci associated with phenotypic trait variation in a collection of individuals. Association mapping uses the natural diversity, which represents many more recombination events occurred in the history of the population, providing better resolution. Nowadays, two association mapping methodologies are in use: candidate gene association, where a good understanding of the biochemistry and genetics of the trait is needed, and whole genome scan, also called genome-wide association (GWA) studies. New genomic advances are providing the higher density of genetic markers required to ensure enough coverage to detect linkage between markers and a causal locus. Also the decrease of sequencing costs (Table 1) has allowed the use of whole genome sequencing in current studies [130].

Nevertheless, association mapping is just rising in model species and major crops. Maize is the most widely studied crop regarding association analysis. Many candidate genes have been successfully associated to morphological or quality traits. As an example, candidate genes Dwarf8, Vgt1 and ZmRap2.7 were successfully associated to flowering time [131]. Other candidate genes have been associated, among others, to forage quality, carotenoid content, oil content and kernel quality [132-135]. GWA studies have been more limited, probably due to the large genome of maize (2300 Mbp) and the great number of markers needed to cover it. The first study identified a fatty acid desaturase gene (fad2) associated with increased oleic acid levels [99]. More recently, other authors found 32 QTLs associated with southern leaf blight disease resistance [100].

Examples of association mapping approaches in other crops are more limited. Studies based on the candidate gene approach have been reported in some crops, like grape, or conifers [102, 106]. However, GWA studies have only been developed either in the model species *A. thaliana* [136] or in major crops such as rice [96], barley [97], or wheat [104]. Some articles also describe successful mapping processes combining classical linkage mapping and association mapping [137]. Although genetic association mapping is in its early steps, it is a promising tool for the dissection of complex traits in crop plants.

Marker Assisted Selection

Marker Assisted Backcross Selection

Marker assisted selection (MAS) is an indirect process where selection is carried out on the basis of a marker instead of the trait itself. The successful application of MAS relies on the tight association between the marker and the major gene or QTL responsible for the trait. As we have described before, the new genomic tools accelerate the identification of markers tightly linked to target genomic regions. On the other hand, the new dense genotyping platforms available today accelerate the genotyping of large amounts of samples during the MAS process in a rapid and economically feasible manner. MAS can take benefit from these technologies, speeding up the release of new varieties.

In spite of the close linkage between the marker and the gene, the possibility of recombination limits the use of MAS. The use of intragenic markers, also called functional markers, can help to overcome this limitation [138]. NGS sequencing projects produce large collections of functional markers. These markers enhance real gene assisted breeding, reducing the possibility of losing the desirable trait due to recombination. This is today feasible in many crop species in which NGS cDNA sequencing is being conducted. Some of these studies perform expression profiling, identifying candidates and associated gene targeted markers.

MAS is also frequently applied to perform background selection in the context of backcrossing programmes. Background selection consists in the identification of plants with lower contents in donor genome to continue the breeding scheme, in order to achieve the recovery of the recipient genome. The use of background markers facilitates the quick recovery of the recurrent parent genome [139]. Background selection is being used extensively in rice breeding. High-density genome maps are being effectively used in background analysis. For example, background selection integrated with foreground selection of bacterial blight resistance (xa13 and Xa21 genes), amylose content (waxy gene) and fertility restorer gene has been performed in order to identify superior lines with maximum recovery of Basmati rice genome along with the quality traits and minimum non-targeted genomic introgressions of the donor chromosomes [140].

In some cases, the problem of recovering the genetic background of the recurrent parent arises because of the linkage drag, that is, the introgression of chromosome regions with deleterious effects which are tightly linked to the gene or QTL of interest. The detection of QTLs responsible of the negative effects and the localization of molecular markers tightly associated to them can be an efficient way to break the genetic drag. A well-known example concerns canola (rapeseed) breeding, which began with the discovery of germplasm with low erucic acid content in seeds of a spring forage cultivar in the 1950's. The problem arose because a high association between low erucic acid content and low seed oil content exists. The recent availability of high-density molecular maps has allowed the detection of several QTLs associated to both traits. Moreover, the identification of molecular markers very tightly linked to the QTLs made possible to break the linkage drag between the low oil content and erucic acid concentration in seeds in the process for breeding new high seed oil content canola cultivars [141].

Frequently, current breeding programmes involve the introgression of more than one gene or QTL controlling traits of interest into one genetic background, in a process that is called pyramiding. The most useful application of MAS in the process of pyramiding is related to the introgression of different genes or QTLs whose effect on the phenotype is undistinguishable. The accumulation of genes from different sources which confer resistance against the same disease is an example, and is indeed one of the most widespread applications of gene pyramiding [142]. The main advantages of recent advances in plant genomics incorporated into gene pyramiding will be related to two different aspects. On one hand, the number of plants to be analyzed in a gene pyramiding programme must be increased as the number of loci of interest is higher, to ensure with a reasonable likelihood that the genotype combining favorable alleles is present in the population [143]. In this sense, the availability of genotyping platforms will provide the possibility to screen larger generations. On the other hand, the efficiency of the process strongly depends on the tightness of the linkage between markers used and the target genes or QTLs. Again, identification of functional markers will circumvent this limitation.

Breeding by Design

The possibility to predict the outcome of a set of crosses on the basis of molecular markers information is known as 'breeding by design' [6]. The process includes three steps: mapping loci involved in all agronomically relevant traits, assessment of the allelic variation at those loci, and, finally, breeding by design. In the method as initially described by Peleman and van der Voort [6], the first step was proposed to be completed by either using mapping populations segregating for the trait of interest or based on a candidate gene approach (mainly exploiting information from model plant species and increasing understanding of gene function). Also linkage disequilibrium (LD) mapping was suggested, focused on the region previously identified as related to the trait ('targeted LD mapping'). Currently, as previously discussed, other possibilities such as GWA studies allow a more efficient way to accomplish this first step, avoiding limitations of biparental populations. The second step of the process consists in the identification of allelic variation for the locus of interest and the assignation of the phenotypic value to each of them. This step cannot be based on biparental populations, given that only two alleles per locus are segregating in this case. The analysis should then include plant materials representing the variability of the species. Genotypic and phenotypic data for each plant are required.

As previously stated, high level of saturation with markers is not the limiting factor in most cases, and so currently the restrictions mostly come from the phenotyping step. Strictly speaking, 'breeding by design' exploits information obtained in the previous steps: once the loci of interest have been mapped, and the contribution of each allelic variant has been determined, crosses can be established to generate superior genotypes which combine all favourable alleles. Application of this breeding strategy has been used for different crops and with different objectives, such as breeding for heading date in rice [144] or seed length in soybean [145]. This procedure has also been used in patent applications; as an example, 'breeding by design' has been reported as part of the development of higher quality maize varieties. However, the most effective application of the 'breeding by design' approach will come from the incorporation of the most advanced genomic tools into the process, which will allow the improvement of the predictions.

Genomic Selection

MAS strategies described so far require the identification of markers associated to the traits of interest. This represents one of the weaknesses of traditional MAS approaches [146]. Nevertheless, MAS can also be applied eluding this step, using an approach known as genomic (or genome-wide) selection (Fig. **11**). The method was first described in 2001 [147], as an attempt to exploit information generated from emerging genotyping technologies. Genomic selection is based on simultaneous estimation of effects on phenotype of all *loci*, haplotypes, and markers available. The difference with other MAS methods relies on the fact that no previous selection of markers with effects on phenotype is developed [148]. Genomic selection requires the availability of phenotypic and genotypic data for the reference population. This data set will allow estimating the parameters for the model, so that the differences at the phenotype level are explained by the markers analysed. Once the model is established, application to breeding populations makes possible to determine the genomic value of each individual, i.e., the expected phenotype based on the genotypic data. The requirement is the availability of enough molecular markers to provide good genome coverage [5, 146].

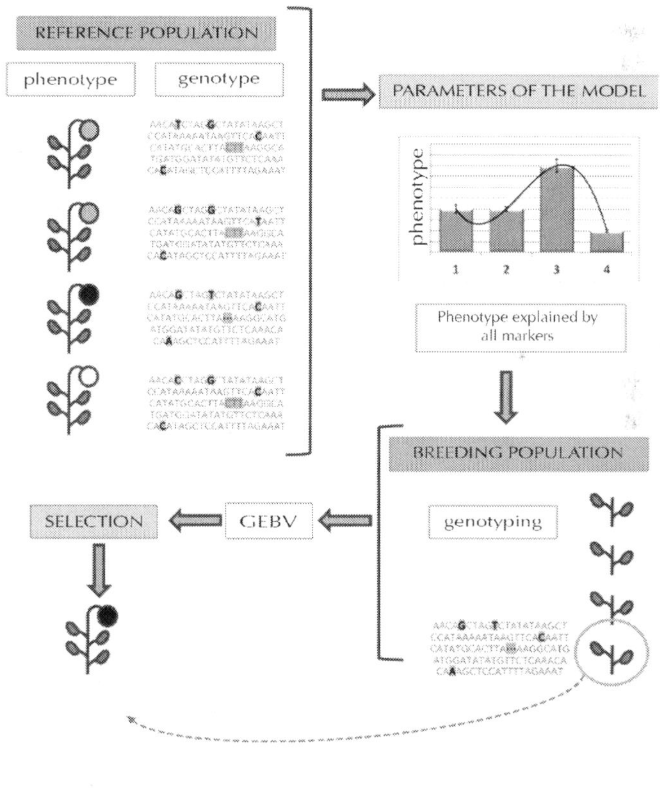

Figure 1: Genomic selection scheme. Information on phenotype and genotype for a reference population allows estimating parameters for the model. This model explains phenotype based on all markers analyzed. The model predicts the phenotype of plants in a breeding popu-

lation on the basis of the genotyping results: this is the genomic estimated breeding value (GEBV), used to select the desired phenotypes.

Simulation studies carried out using maize proved the usefulness of genomic selection applied to an initial cross between an adapted line and exotic germplasm. With 512 markers and a reference population of 288 F_2plants evaluated in six different environments, it was possible to obtain good selection response after 7-8 generations. [149]. Also with maize, simulations showed that response to selection was 18 to 43% larger for genomic selection than for marker assisted recurrent selection [150]. Response obtained when using genomic selection can be lower than response by phenotypic selection. However, the reduction in cycle length due to early MAS results in an increase of gain per time unit. This reduction is even more accused for species with a long generation interval, such as tree species [148].

The availability of phenotypic databases for different crops has allowed the comparison of predictions about the genotypic value obtained using genomic selection with the true genotypic value as shown by the phenotypic manifestation of the trait. In a study developed with phenotypic and genotypic data from *Arabidopsis*, maize and barley, results obtained were more accurate when genome-wide selection was carried out, if compared with results derived of previous selection of markers with effects on the phenotype [151].

Although when applying genomic selection there is no need to previously identify QTLs controlling a certain trait, the utilization of this approach allows detecting the chromosome regions involved in a given trait, as markers with greater effect on the phenotype will indicate the presence of a QTL for this trait [152]. Some studies go one step farther and propose the application of MAS prior to phenotyping. This approach involves the use of *prior indices*, i.e., marker selection indices which have been constructed from a given phenotyped and genotyped population and are applied to different populations which have not been phenotyped [129]. The decrease in the costs of genotyping provides the appropriate scenario for this strategy to become cost-effective.

In any case, even the identification of the QTLs responsible for a certain trait does not imply the identification of the gene or genes controlling the trait itself or the understanding of the mode of action. Models applied in genomic selection are useful to predict breeding values and, in some cases, detect regions associated to a trait, but further work is necessary from this point to identify the gene or genes responsible for the phenotypic variability observed. From plant breeders' perspective, the availability of molecular markers which allow MAS to be applied is generally sufficient. However, development of the new high throughput -omics technologies has provided breeders with new strategies to search for candidate genes, mainly based on microarray for differential gene expression, being the possibility to explore more genes the most important advantage. Future exploitation of these strategies could facilitate the identification of candidate genes underlying the traits of interest and make MAS more efficient.

CONCLUSIONS

For some major crops the pace experimented for genetic gains in yield and other complex traits in the 20thcentury will be difficult to be maintained if only existing pre-genomics technologies are used [153]. However, plant breeding is a dynamic science and, fortunately, genomics resources and tools are already available and are helping to give another quantitative leap in plant breeding. In this respect many advances are already taking place, and the superdomestication, i.e., "the processes that lead to a domesticate with dramatically increased yield that could not be selected in natural environments from naturally occurring variation without recourse to new technologies" [10], will require the combination of conventional breeding with crop genomics. Also, genomic tools and approaches will help conventional breeding in achieving important advances in the breeding of crops that from the point of view of genetic improvement have remained either orphaned or neglected [8]. Therefore, while conventional pre-genomics plant breeding has been, is, and will be successful at improving our crops, the application of genomic tools and resources to practical plant breeding will push forward the genetic gains obtained by breeding programmes. New genomic advances, many of which are already being developed, will make easier for breeders to obtain new cultivars with improved characteristics, either by facilitating selection or by improving the variation available for breeders by using precision breeding approaches. In particular, the present and new genomics tools are of great value for the genetic dissection and breeding of complex traits.

ACKNOWLEDGEMENTS

This contribution has been partially funded by the Ministerio de Ciencia y Tecnología (AGL2008-05114, AGL2009-07257, and PIM2010PKB-00691), and Instituto Nacional de Investigación y Tecnología Agraria y Agroalimentaria (RTA2008-00035-C02-02 and RTA2011-00044-C02-2) and FEDER funds (EU).

ABBREVIATIONS

AFLP = Amplified Fragment Length Polymorphism
BED= Browser Extensible Data
BSA = Bulked Segregant Analysis
CSSL = Chromosome Segment Substitution Line
DArT = Diversity Arrays Technology
EcoTILLING = Ecotype TILLING
EST = Expressed Sequence Tag
GFF = General Feature Format
GWA = Genome Wide Association
LD = Linkage Disequilibrium
MAF = Minor Allele Frequency
MAS = Marker Assisted Selection
MPSS = Massively Parallel Signature Sequencing
NGS = Next Generation Sequencing
NIL = Near Isogenic Line
QTL = Quantitative Trait Locus
RAD = Restriction-Site Associated DNA
RIL = Recombinant Inbred Line
RNA-seq = Sequencing of RNA Transcripts
SAGE = Serial Analysis of Gene Expression
SAM = Sequence Alignment and Modelling
SPF = Single-Feature Polymorphism

SNP = Single Nucleotide Polymorphism

SSR = Simple Sequence Repeat

TILLING = Targeting Induced Local Lesions on Genomes

VCF = Variant Call Format

REFERENCES

1. Fedoroff NV. The past, present and future of crop genetic modification. New Biotechnol. 2010;27:461–465.
2. Tester M, Langridge P. Breeding technologies to increase crop production in a changing world. Science.2010;327:818–822.
3. Varshney RK, Tuberosa R. Genomics-assisted crop improvement vol. 1: Genomics approaches and platforms. New York: Springer; 2007a.
4. Varshney RK, Tuberosa R. Genomics-assisted crop improvement vol. 2: Genomics applications in crops.New York: Springer; 2007b.
5. Lorenz AJ, Chao S, Asoro FG, Heffner EL, Hayashi T, Iwata H, Smith KP, Sorrells MK, Jannink JL. Genomic selection in plant breeding: knowledge and prospects. Adv. Agron. 2011;110:77–123.
6. Peleman JD, van der Voort JR. Breeding by design. Trends Plant Sci. 2003;8:330–334.
7. Collard BCY, Mackill DJ. Marker-assisted selection: an approach for precision plant breeding in the twenty-first century. Phil. Transact. Royal Soc. B. 2008;363:557–572.
8. Varshney RK, Glaszmann JC, Leung H, Ribaut JM. More genomic resources for less-studied crops.Trends Biotechnol. 2010;28:452–460.
9. Morgante M, Salamini F. From plant genomics to breeding practice. Curr. Opin. Biotechnol.2003;14:214–219.
10. Vaughan DA, Balász E, Heslop-Harrison JS. From crop domestication to super-domestication. Ann. Bot.2007;100:893–901.

11. The Arabidopsis Genome Initiative. Analysis of the genome sequence of the flowering plant Arabidopsis thaliana. Nature. 2010;408:796–815.

12. International Rice Genome Sequencing Project. The map-based sequence of the rice genome. Nature.2005;436:793–800.

13. Schnable PS, Ware D, Fulton RS, Stein JC, Wei F, Pasternak S, Liang C, Zhang J, Fulton L, Graves TA, Minx P, Reily AD, Courtney L, Kruchowski SS, Tomlinson C, Strong C, Delehaunty K, Fronick C, Courtney B, Rock SM, Belter E, Du F, Kim K, Abbott RM, Cotton M, Levy A, Marchetto P, Ochoa K, Jackson SM, Gillam B, Chen W, Yan L, Higginbotham J, Cardenas M, Waligorski J, Applebaum E, Phelps L, Falcone J, Kanchi K, Thane T, Scimone A, Thane N, Henke J, Wang T, Ruppert J, Shah N, Rotter K, Hodges J, Ingenthron E, Cordes M, Kohlberg S, Sgro J, Delgado B, Mead K, Chinwalla A, Leonard S, Crouse K, Collura K, Kudrna D, Currie J, He R, Angelova A, Rajasekar S, Mueller T, Lomeli R, Scara G, Ko A, Delaney K, Wissotski M, Lopez G, Campos D, Braidotti M, Ashley E, Golser W, Kim H, Lee S, Lin J, Dujmic Z, Kim W, Talag J, Zuccolo A, Fan C, Sebastian A, Kramer M, Spiegel L, Nascimento L, Zutavern T, Miller B, Ambroise C, Muller S, Spooner W, Narechania A, Ren L, Wei S, Kumari S, Faga B, Levy MJ, McMahan L, Van Buren P, Vaughn MW, Ying K, Yeh CT, Emrich SJ, Jia Y, Kalyanaraman A, Hsia AP, Barbazuk WB, Baucom RS, Brutnell TP, Carpita NC, Chaparro C, Chia JM, Deragon JM, Estill JC, Fu Y, Jeddeloh JA, Han Y, Lee H, Li P, Lisch DR, Liu S, Liu Z, Nagel DH, McCann MC, SanMiguel P, Myers AM, Nettleton D, Nguyen J, Penning BW, Ponnala L, Schneider KL, Schwartz DC, Sharma A, Soderlund C, Springer NM, Sun Q, Wang H, Waterman M, Westerman R, Wolfgruber TK, Yang L, Yu Y, Zhang L, Zhou S, Zhu Q, Bennetzen JL, Dawe RK, Jiang J, Jiang N, Presting GG, Wessler SR, Aluru S, Martienssen RA, Clifton SW, McCombie WR, Wing RA, Wilson RK. The B73 maize genome: complexity, diversity, and dynamics. Science. 2009;326:1112–1115.

14. Paterson AH, Bowers JE, Bruggmann R, Dubchak I, Grimwood J, Gundlach H, Haberer G, Hellsten U, Mitros T, Poliakov A, Schmutz J, Spannagl M, Tang H, Wang X, Wicker T, Bharti AK, Chapman J, Feltus FA, Gowik U, Grigoriev IV, Lyons E, Maher CA, Martis M, Narechania A, Otillar RP, Penning BW, Salamov AA, Wang Y, Zhang L, Carpita NC, Freeling M, Gingle AR, Hash

CT, Keller B, Klein P, Kresovich S, McCann MC, Ming R, Peterson DG, Rahman M, Ware D, Westhoff P, Mayer KFX, Messing J, Rokhsar DS. The *Sorghum bicolor* genome and the diversification of grasses. Nature. 2009;457:551–556.

15. Tuskan GA, Difazio S, Jansson S, Bohlmann J, Grigoriev I, Hellsten U, Putnam N, Ralph S, Rombauts S, Salamov A, Schein J, Sterck L, Aerts A, Bhalerao RR, Bhalerao RP, Blaudez D, Boerjan W, Brun A, Brunner A, Busov V, Campbell M, Carlson J, Chalot M, Chapman J, Chen GL, Cooper D, Coutinho PM, Couturier J, Covert S, Cronk Q, Cunningham R, Davis J, Degroeve S, Déjardin A, Depamphilis C, Detter J, Dirks B, Dubchak I, Duplessis S, Ehlting J, Ellis B, Gendler K, Goodstein D, Gribskov M, Grimwood J, Groover A, Gunter L, Hamberger B, Heinze B, Helariutta Y, Henrissat B, Holligan D, Holt R, Huang W, Islam-Faridi N, Jones S, Jones-Rhoades M, Jorgensen R, Joshi C, Kangasjärvi J, Karlsson J, Kelleher C, Kirkpatrick R, Kirst M, Kohler A, Kalluri U, Larimer F, Leebens-Mack J, Leplé JC, Locascio P, Lou Y, Lucas S, Martin F, Montanini B, Napoli C, Nelson DR, Nelson C, Nieminen K, Nilsson O, Pereda V, Peter G, Philippe R, Pilate G, Poliakov A, Razumovskaya J, Richardson P, Rinaldi C, Ritland K, Rouzé P, Ryaboy D, Schmutz J, Schrader J, Segerman B, Shin H, Siddiqui A, Sterky F, Terry A, Tsai CJ, Uberbacher E, Unneberg P, Vahala J, Wall K, Wessler S, Yang G, Yin T, Douglas C, Marra M, Sandberg G, Van de Peer Y, Rokhsar D. The genome of black cottonwood, *Populus trichocarpa* (Torr & Gray) Science.2006;313:1596–1604.

16. Jaillon O, Aury JM, Noel B, Policriti A, Clepet C, Casagrande A, Choisne N, Aubourg S, Vitulo N, Jubin C, Vezzi A, Legeai F, Hugueney P, Dasilva C, Horner D, Mica E, Jublot D, Poulain J, Bruyère C, Billault A, Segurens B, Gouyvenoux M, Ugarte E, Cattonaro F, Anthouard V, Vico V, Del Fabbro C, Alaux M, Di Gaspero G, Dumas V, Felice N, Paillard S, Juman I, Moroldo M, Scalabrin S, Canaguier A, Le Clainche I, Malacrida G, Durand E, Pesole G, Laucou V, Chatelet P, Merdinoglu D, Delledonne M, Pezzotti M, Lecharny A, Scarpelli C, Artiguenave F, Pè ME, Valle G, Morgante M, Caboche M, Adam-Blondon AF, Weissenbach J, Quétier F, Wincker P. French-Italian Public Consortium for Grapevine Genome Characterization. The grapevine genome sequence suggests ancestral hexaploidization in major angiosperm phyla. Nature. 2007;449:463–467.

17. Ming R, Hou S, Feng Y, Yu O, Dionne-Laporte A, Saw JH, Senin P, Wang W, Ly BV, Lewis KLT, Salzberg SL, Feng L, Jones MR, Skelton RL, Murray JE, Chen C, Qian W, Shen J, Du P, Moriah Eustice ME, Tong E, Tang H, Lyons E, Paull RE, Michael TP, Wall K, Rice DW, Albert H, Wang ML, Zhu YJ, Schatz M, Nagarajan N, Acob RA, Guan P, Blas A, Wai CM, Ackerman CM, Ren Y, Liu C, Wang J, Wang J, Na JK, Shakirov EV, Brian Haas B, Thimmapuram J, Nelson D, Wang X, Bowers JE, Gschwend AR, Delcher AL, Singh R, Suzuki JY, Tripathi S, Neupane K, Wei H, Irikura B, Paidi M, Jiang N, Zhang W, Presting G, Windsor A, Navajas-Pérez R, Torres MJ, Feltus FA, Porter B, Li Y, Burroughs AM, Luo MC, Liu L, Christopher DA, Mount SM, Moore PH, Sugimura T, Jiang J, Schuler MA, Friedman V, Mitchell-Olds T, Shippen DE, dePamphilis CW, Palmer JD, Freeling M, Paterson AH, Gonsalves D, Wang L, Alam M. The draft genome of the transgenic tropical fruit tree papaya (*Carica papaya* Linnaeus) Nature .2008;452:991–996.

18. Schmutz J, Cannon SB, Schlueter J, Ma J, Mitros T, Nelson W, Hyten DL, Song Q, Thelen JJ, Cheng J, Xu D, Hellsten U, May GD, Yu Y, Sakurai T, Umezawa T, Bhattacharyya MK, Sandhu D, Valliyodan B, Lindquist E, Peto M, Grant D, Shu S, Goodstein D, Barry K, Futrell-Griggs M, Abernathy B, Du J, Tian Z, Zhu L, Gill N, Joshi T, Libault M, Sethuraman A, Zhang XC, Shinozaki K, Nguyen HT, Wing RA, Cregan P, Specht J, Grimwood J, Rokhsar D, Stacey G, Shoemaker RC, Jackson SA. Genome sequence of the palaeopolyploid soybean. Nature. 2010;463:178–183.

19. Metzker M. Sequencing technologies-the next generation. Nat. Rev. Genet. 2010;11:31.

20. Wang Z, Fang B, Jingyi C, Zhang X, Luo X, Huang L, Chen X, Li Y. De novo assembly and characterization of root transcriptome using Illumina paired-end sequencing and development of cSSR markers in sweetpotato (*Ipomoea batatas*) BMC Genomics. 2011;11:726.

21. Blanca J, Cañizares J, Roig C, Ziarsolo P, Nuez F, Picó B. Transcriptome characterization and high throughput SSRs and SNPs discovery in *Cucurbita pepo* (Cucurbitaceae) BMC Genomics. 2011;12:104.

22. Dutta S, Kumawat G, Singh BP, Gupta DK, Singh S, Dogra V, Gaikwad K, Sharma TR, Raje RS, Bandhopadhya TK, Datta S,

Singh MN, Bashasab F, Kulwal P, Wanjari KB, Varshney RK, Cook DR, Singh NK. Development of genic-SSR markers by deep transcriptome sequencing in pigeonpea [*Cajanus cajan* (L.) Millspaugh] BMC Plant Biol. 2011;11:17.

23. Logacheva MD, Kasianov AS, Vinogradov DV, Samigullin TH, Gelfand MS, Makeev VJ, Penin AA.*De novo* sequencing and characterization of floral transcriptome in two species of buckwheat (*Fagopyrum*)BMC Genomics. 2011;12:30.
24. Li H, Handsaker B, Wysoker A, Fennell T, Ruan J, Homer N, Marth G, Abecasis G, Durbin R. Genome Project Data P: The Sequence Alignment/Map format and SAMtools. Bioinformatics. 2009;25:2078–2079.
25. Pop M, Salzberg SL. Bioinformatics challenges of new sequencing technology. Trends Genet.2008;24:142–149.
26. Horner DS, Pavesi G, Castrignanò T, Meo PDOD, Liuni S, Sammeth M, Picardi E, Presole G. Bioinformatics approaches for genomics and post genomics applications of next-generation sequencing.Brief. Bioinform. 2009;2:181–197.
27. Li H, Horner DS. A survey of sequence alignment algorithms for next-generation sequencing. Brief. Bioinform. 2010;2:473–483.
28. Wang L, Li P, Brutnell TP. Exploring plant transcriptomes using ultra high-throughput sequencing.Brief. Funct. Genomics. 2010;9:118–128.
29. VanGuilder HD, Vrana KE, Freeman WM. Twenty-five years of quantitative PCR for gene expression analysis. Biotechniques. 2008;44:619–626.
30. Bachem CW, van der Hoeven RS, de Bruijn SM, Vreugdenhil D, Zabeau M, Visser RG. Visualization of differential gene expression using a novel method of RNA fingerprinting based on AFLP: analysis of gene expression during potato tuber development. Plant J. 1996;9:745–753.
31. Anisimov SV. Serial Analysis of Gene Expression (SAGE): 13 years of application in research. Curr. Pharm. Biotechnol. 2008;9:338–350.
32. Reinartz J, Bruyns E, Lin JZ, Burcham T, Brenner S, Bowen B, Kramer M, Woychik R. Massively parallel signature sequencing (MPSS) as a tool for in-depth quantitative gene expression profiling

in all organisms. Brief. Funct. Genomics Proteomics. 2002;1:95–104.

33. Schena M, Shalon D, Davis RW, Brown PO. Quantitative monitoring of gene expression patterns with a complementary DNA microarray. Science. 1995;270:467–470.
34. Al-Shahrour F, Minguez P, Vaquerizas JM, Conde L, Dopazo J. BABELOMICS: a suite of web tools for functional annotation and analysis of groups of genes in high-throughput experiments. Nucleic Acids Res. 2005;33:460–464.
35. Saeed AI, Sharov V, White J, Li J, Liang W, Bhagabati N, Braisted J, Klapa M, Currier T, Thiagarajan M, Sturn A, Snuffin M, Rezantsev A, Popov D, Ryltsov A, Kostukovich E, Borisovsky I, Liu Z, Vinsavich A, Trush V, Quackenbush J. TM4: a free, open-source system for microarray data management and analysis. Biotechniques. 2003;34:374–378.
36. Zimmermann P, Laule O, Schmitz J, Hruz T, Bleuler S, Gruissem W. Genevestigator transcriptome meta-analysis and biomarker search using rice and barley gene expression databases. Mol. Plant.2008;1:851–857.
37. Schmid M, Davison TS, Henz SR, Pape UJ, Demar M, Vingron M, Scholkopf B, Weigel D, Lohmann JU. A gene expression map of Arabidopsis thaliana development. Nat. Genet. 2005;37:501–506.
38. Barrett T, Troup DB, Wilhite SE, Ledoux P, Evangelista C, Kim IF, Tomashevsky M, Marshall KA, Phillippy KH, Sherman PM, Muertter RN, Holko M, Ayanbule O, Yefanov A, Soboleva A. NCBI GEO: archive for functional genomics data sets -10 years on. Nucleic Acids Res. 2011;39:1005–1010.
39. Parkinson H, Sarkans U, Kolesnikov N, Abeygunawardena N, Burdett T, Dylag M, Emam I, Farne A, Hastings E, Holloway E, Kurbatova N, Lukk M, Malone J, Mani R, Pilicheva E, Rustici G, Sharma A, Williams E, Adamusiak T, Brandizi M, Sklyar N, Brazma A. ArrayExpress update - an archive of microarray and high-throughput sequencing-based functional genomics experiments. Nucleic Acids Res.2011;39:1002–1004.
40. Pascual L, Blanca JM, Canizares J, Nuez F. Analysis of gene expression during the fruit set of tomato: A comparative approach. Plant Sci. 2007;173:609–620.

41. Marioni JC, Mason CE, Mane SM, Stephens M, Gilad Y. RNA-seq: an assessment of technical reproducibility and comparison with gene expression arrays. Genome Res. 2008;18:1509–1517.

42. Stiglic G, Bajgot M, Kokol P. Gene set enrichment meta-learning analysis: next-generation sequencing versus microarrays. BMC Bioinformatics. 2010;11:176.

43. Cloonan N, Forrest AR, Kolle G, Gardiner BB, Faulkner GJ, Brown MK, Taylor DF, Steptoe AL, Wani S, Bethel G, Robertson AJ, Perkins AC, Bruce SJ, Lee CC, Ranade SS, Peckham HE, Manning JM, McKernan KJ, Grimmond SM. Stem cell transcriptome profiling via massive-scale mRNA sequencing. Nat. Methods. 2008;5:613–619.

44. Alagna F, D›Agostino N, Torchia L, Servili M, Rao R, Pietrella M, Giuliano G, Chiusano ML, Baldoni L, Perrotta G. Comparative 454 pyrosequencing of transcripts from two olive genotypes during fruit development. BMC Genomics. 2009;10:399.

45. Zenoni S, Ferrarini A, Giacomelli E, Xumerle E, Fasoli M, Malerba G, Bellin D, Pezzotti M, Delledonne M. Characterization of transcriptional complexity during berry development in *Vitis vinifera* using RNA-Seq.Plant Phisiol . 2010;152:1787–1795.

46. Wang Y, Zhang H, Li H, Miao X. Second-generation sequencing supply an effective way to screen RNAi targets in large scale for potential application in pest insect control. PLoS One. 2011;6(4):e18644.

47. Shi CY, Yang H, Wei CL, Yu O, Zhang ZZ, Jiang CJ, Sun J, Li YY, Chen Q, Xia T, Wan XC. Deep sequencing of the *Camellia sinensis* transcriptome revealed candidate genes for major metabolic pathways of tea-specific compounds. BMC Genomics. 2011;12:131.

48. Gepts P. Plant genetic resources conservation and utilization: the accomplishments and future of a societal insurance policy. Crop Sci. 2006;46:2278–2292.

49. Parry MAJ, Madgwick PJ, Bayon C, Tearall K, Hernandez-Lopez A, Baudo M, Rakszegi M, Hamada W, Al-Yassin A, Ouabbou H, Labhilili M, Phillips AL. Mutation discovery for crop improvement. J. Exp. Bot. 2009;60:2817–2825.

50. Till BJ, Reynolds SH, Greene EA, Codomo CA, Enns LC, Johnson JE, Burtner C, Odden AR, Young K, Taylor NE, Henikoff JG, Comai L, Henikoff S. Large-scale discovery of induced point mutations with high- throughput TILLING. Genome Res. 2003;13:524–530.
51. Comai L, Young K, Till BJ, Reynolds SH, Greene EA, Codomo CA, Enns LC, Johnson JE, Burtner C, Odden AR, Henikoff S. Efficient discovery of DNA polymorphisms in natural populations by EcoTILLING. Plant J . 2004;37:778–786.
52. Till BJ, Burtner C, Comai L, Henikoff S. Mismatch cleavage by single-strand specific nucleases. Nucleic Acids Res. 2004;32:2632–2641.
53. Triques K, Piednoir E, Dalmais M, Schmidt J, Le Signor C, Sharkey M, Caboche M, Sturbois B, Bendahmane A. Mutation detection using ENDO1: Application to disease diagnostics in humans and TILLING and EcoTILLING in plants. BMC Mol. Biol. 2008;9:9.
54. Barkley NA, Wang ML. Application of TILLING and EcoTILLING as reverse genetic approaches to elucidate the function of genes in plants and animals. Curr. Genomics. 2008;9:212–226.
55. Colbert T, Till BJ, Tompa R, Reynolds S, Steine MN, Yeung AT, McCallum CM, Comai L, Henikoff S. High-throughput screening for induced point mutations. Plant Physiol. 2001;126:480–484.
56. Perry JA, Wang TL, Welham TJ, Gardner S, Pike JM, Yoshida S, Parniske M. A TILLING reverse genetics tool and a web-accessible collection of mutants of the legume *Lotus japonicus*. Plant Physiol.2003;131:866–871.
57. Caldwell DG, McCallum N, Shaw P, Muehlbauer GJ, Marshall DF, Waugh R. A structured mutant population for forward and reverse genetics in barley (*Hordeum vulgare* L) Plant J . 2004;40:143–150.
58. Weil CF, Monde RA. Getting to the point - mutations in maize. Crop Sci. 2007. pp. S60–S67.
59. Triques K, Sturbois B, Gallais S, Dalmais M, Chauvin S, Clepet C, Aubourg S, Rameau C, Caboche M, Bendahmane A. Characterization of *Arabidopsis thaliana* mismatch specific endonucleases: application to mutation discovery by TILLING in pea. Plant J. 2007;51:1116–1125.

60. Dahmani-Mardas F, Troadec C, Boualem A, Lévêque S, Alsadon AA, Aldoss AA, Dogimont C, Bendahmane A. Engineering melon plants with improved fruit shelf life using the TILLING approach. PLoS One. 2010;5(12):e15776.
61. Kadaru SB, Yadav AS, Fjellstrom RG, Oard JH. Alternative EcoTILLING protocol for rapid, cost-effective single-nucleotide polymorphism discovery and genotyping in rice (*Oryza sativa* L.) Plant Mol. Biol. Reporter. 2006;24:3–22.
62. Mejlhede N, Kyjovska Z, Backes G, Burhenne K, Rasmussen SK, Jahoor A. EcoTILLING for the identification of allelic variation in the powdery mildew resistance genes *mlo* and *Mla* of barley. Plant Breed.2006;125:461–467.
63. Wang J, Sun JZ, Liu DC, Yang WL, Wang DW, Tong YP, Zhang AM. Analysis of *Pina* and *Pinb*alleles in the micro-core collections of Chinese wheat germplasm by EcoTILLING and identification of a novel *Pinb* allele. J Cereal Sci. 2008;48:836–842.
64. Ramos ML, Huntley JJ, Maleki SJ, Ozias-Akins P. Identification and characterization of a hypoallergenic ortholog of Ara h 2.01. Plant Mol Biol. 2009;69:325–335.
65. Galeano CH, Gomez M, Rodriguez LM, Blair MW. CEL I Nuclease Digestion for SNP Discovery and Marker Development in Common Bean (*Phaseolus vulgaris* L.) Crop Sci . 2009;49:381–394.
66. Wang J, Sun JZ, Liu DC, Yang WL, Wang DW, Tong YP, Zhang AM. Analysis of *Pina* and *Pinb*alleles in the micro-core collections of Chinese wheat germplasm by EcoTILLING and identification of a novel *Pinb* allele. J. Cereal Sci. 2008;48:836–842.
67. Ramos ML, Huntley JJ, Maleki SJ, Ozias-Akins P. Identification and characterization of a hypoallergenic ortholog of Ara h 2.01. Plant Mol. Biol. 2009;69:325–335.
68. Piron F, Nicolaï M, Minoïa S, Piednoir E, Moretti A, Salgues A, Zamir D, Caranta C, Bendahmane A. An induced mutation in tomato eIF4E leads to immunity to two potyviruses. PLoS One. 2010;5(6):e11313.
69. Ibiza VP, Cañizares J, Nuez F. EcoTILLING in Capsicum species: searching for new virus resistances.BMC Genomics. 2010;11:631.

70. Deschamps S, Campbell MA. Utilization of next-generation sequencing platforms in plant genomics and genetic variant discovery. Mol. Breed. 2010;25:553–570.
71. Ganal MW, Altmann T, Röder MS. SNP identification in crop plants. Curr. Opin. Plant Biol.2009;12:211–217.
72. Weigel D, Mott R. The 1001 genomes project for *Arabidopsis thaliana*. Genome Biol. 2009;10:107.
73. Gore MA, Wright MH, Ersoz ES, Bouffard P, Szekeres ES, Jarvie TP, Hurwitz BL, Narechania A, Harkins TT, Grills GS, Ware DH, Buckler ES. Large-scale discovery of gene-enriched SNPs. Plant Genome. 2009;2:121–133.
74. Myles S, Chia JM, Hurwitz B, Simon C, Zhong GY, Buckler E, Ware D. Rapid genomic characterization of the genus *Vitis*. PLoS One. 2010;5(1):e8219.
75. McCouch SR, Zhao K, Wright M, Tung CW, Ebana K, Thomson M, Reynolds A, Wang D, DeClerck G, Ali ML, McClung A, Eizenga G, Bustamante C. Development of genome-wide SNP assays for rice.Breed. Sci. 2010;60:524–535.
76. Wu X, Ren C, Joshi T, Vuong T, Xu D, Nguyen HT. SNP discovery by high-throughput sequencing in soybean. BMC Genomics. 2010;11:469.
77. Deschamps S, La Rota M, Ratashak JP, Biddle P, Thureen D, Farmer A, Luck S, Beatty M, Nagasawa N, Michael L, Llaca V, Sakai H, May G, Lightner J, Campbell MA. Rapid genome-wide single nucleotide polymorphism discovery in soybean and rice via deep resequencing of reduced representation libraries with the Illumina genome analyzer. Plant Genome. 2010;3:1.
78. Bundock PC, Eliott FG, Ablett G, Benson AD, Casu RE, Aitken KS, Henry RJ. Targeted single nucleotide polymorphism (SNP) discovery in a highly polyploid plant species using 454 sequencing. Plant Biotechnol. J. 2009;7:347–354.
79. TrickM, LongY, MengJ, BancroftI. Singlenucleotidepolymorphism (SNP) discovery in the polyploid*Brassica napus* using Solexa transcriptome sequencing. Plant Biotechnol. J. 2009;7:334–346.
80. Oliver RE, Lazo GR, Lutz JD, Rubenfield MJ, Tinker NA, Anderson JM, Morehead NHW, Adhikary D, Jellen EN, Maughan PJ, Brown Guedira GL, Chao S, Beattie AD, Carson ML, Rines HW, Obert

DE, Bonman JM, Jackson EW. Model SNP development for complex genomes based on hexaploid oat using high-throughput 454 sequencing technology. BMC Genomics. 2011;12:77.

81. Shen Y, Wan Z, Coarfa C, Drabek R, Chen L, Ostrowski EA, Liu Y, Weinstock GM, Wheeler DA, Gibbs RA, Yu F. A SNP discovery method to assess variant allele probability from next-generation resequencing data. Genome Res. 2010;20:273–280.

82. Maughan PJ, Smith SM, Fairbanks DJ, Jellen EN. Development, characterization, and linkage mapping of single nucleotide polymorphisms in the grain amaranths (*Amaranthus* sp.) Plant Genome. 2011;4:92–101.

83. Hyten DL, Cannon SB, Song Q, Weeks N, Fickus EW, Shoemaker RC, Specht JE, Farmer AD, May GD, Cregan PB. High-throughput SNP discovery through deep resequencing of a reduced representation library to anchor and orient scaffolds in the soybean whole genome sequence. BMC Genomics. 2010;11:38.

84. You FM, Huo N, Deal KR, Gu YQ, Luo MC, McGuire PE, Dvorak J, Anderson OD. Annotation-based genome-wide SNP discovery in the large and complex *Aegilops tauschii* genome using next-generation sequencing without a reference genome sequence. BMC Genomics. 2011;12:59.

85. Barbazuk WB, Emrich SJ, Hsin D, Chen L, Li P, Schnable PS. SNP discovery via 454 transcriptome sequencing. Plant J. 2007;51:910–918.

86. Novaes E, Drost DR, Farmerie WG, Pappas GJ, Jr, Grattapaglia D, Sederoff RR, Kirst M. High-throughput gene and SNP discovery in *Eucalyptus grandis*, an uncharacterized genome. BMC Genomics.2008;9:312.

87. Ueno S, Le Provost G, Léger V, Klopp C, Noirot C, Frigerio JM, Salin F, Salse J, Abrouk M, Murat F, Brendel O, Derory J, Abadie P, Léger P, Cabane C, Barré A, de Daruvar A, Couloux A, Wincker P, Reviron MP, Kremer A, Plomion C. Bioinformatic analysis of ESTs collected by Sanger and pyrosequencing methods for a keystone forest tree species: oak. BMC Genomics. 2010;11:650.

88. Vidal RO, Mondego JM, Pot D, Ambrósio AB, Andrade AC, Pereira LF, Colombo CA, Vieira LG, Carazzolle MF, Pereira GA. A high-throughput data mining of single nucleotide polymorphisms in *Coffea*species expressed sequence tags suggests differential

homeologous gene expression in the allotetraploid*Coffea arabica*. Plant Physiol. 2010;154:1053–1066.

89. Schafleitner R, Tincopa LR, Palomino O, Rossel G, Robles RF, Alagon R, Rivera C, Quispe C, Rojas L, Pacheco JA, Solis J, Cerna D, Kim JY, Hou J, Simon R. A sweet potato gene index established by de novo assembly of pyrosequencing and Sanger sequences and mining for gene-based microsatellite markers. BMC Genomics. 2010;11:604.
90. Garg R, Patel RK, Tyagi AK, Jain M. De novo assembly of chickpea transcriptome using short reads for gene discovery and marker identification. DNA Res. 2011;18:53–63.
91. Shirasawa K, Isobe S, Hirakawa H, Asamizu E, Fukuoka H, Just D, Rothan C, Sasamoto S, Fujishiro T, Kishida Y, Kohara M, Tsuruoka H, Wada T, Nakamura Y, Sato S, Tabata S. SNP discovery and linkage map construction in cultivated tomato. DNA Res. 2010;17:381–391.
92. Guo S, Zheng Y, Joung JG, Liu S, Zhang Z, Crasta OR, Sobral BW, Xu Y, Huang S, Fei Z. Transcriptome sequencing and comparative analysis of cucumber flowers with different sex types. BMC Genomics. 2010;11:384.
93. Blanca J, Cañizares J, Ziarsolo P, Esteras C, Mir G, Nuez F, García-Mas J, Picó B. Melon transcriptome characterization. SSRs and SNPs discovery for high throughput genotyping across the species. Plant Genome. 2011;4:118–131.
94. Narina SS, Buyyarapu R, Kottapalli KR, Sartie AM, Ali MI, Robert A, Hodeba MJD, Sayre BL, Scheffler BE. Generation and analysis of expressed sequence tags (ESTs) for marker development in yam (*Dioscorea alata* L.) BMC Genomics. 2011;12:100.
95. Yang SS, Tu ZJ, Cheung F, Xu WW, Lamb JFS, Jung HJG, Vance CP, Gronwald JW. Using RNA-Seq for gene identification, polymorphism detection and transcript profiling in two alfalfa genotypes with divergent dell wall composition in stems. BMC Genomics. 2011;12:199.
96. Huang XH, Wei XH, Sang T, Zhao QA, Feng Q, Zhao Y, Li CY, Zhu CR, Lu TT, Zhang ZW, Li M, Fan DL, Guo YL, Wang A, Wang L, Deng LW, Li WJ, Lu YQ, Weng QJ, Liu KY, Huang T, Zhou TY, Jing YF, Li W, Lin Z, Buckler ES, Qian QA, Zhang QF, Li JY, Han B.

Genome-wide association studies of 14 agronomic traits in rice landraces. Nat. Genet. 2010;42:961–976.

97. Massman J, Cooper B, Horsley R, Neate S, Dill-Macky R, Chao S, Dong Y, Schwarz P, Muehlbauer GJ, Smith KP. Genome-wide association mapping of Fusarium head blight resistance in contemporary barley breeding germplasm. Mol. Breed. 2011;27:439–454.

98. Close TJ, Bhat PR, Lonardi S, Wu YH, Rostoks N, Ramsay L, Druka A, Stein N, Svensson JT, Wanamaker S, Bozdag S, Roose ML, Moscou MJ, Chao SAM, Varshney RK, Szucs P, Sato K, Hayes PM, Matthews DE, Kleinhofs A, Muehlbauer GJ, DeYoung J, Marshall DF, Madishetty K, Fenton RD, Condamine P, Graner A, Waugh R. Development and implementation of high-throughput SNP genotyping in barley. BMC Genomics. 2009;10:582.

99. Belo A, Zheng PZ, Luck S, Shen B, Meyer DJ, Li BL, Tingey S, Rafalski A. Whole genome scan detects an allelic variant of fad2 associated with increased oleic acid levels in maize. Mol. Genet. Genomics.2008;279:1–10.

100. Kump KL, Bradbury PJ, Wisser RJ, Buckler ES, Belcher AR, Oropeza-Rosas MA, Zwonitzer JC, Kresovich S, McMullen MD, Ware D, Balint-Kurti PJ, Holland JB. Genome-wide association study of quantitative resistance to southern leaf blight in the maize nested association mapping population. Nat. Genet.2011;43:163–168.

101. Yan JB, Yang XH, Shah T, Sanchez-Villeda H, Li JS, Warburton M, Zhou Y, Crouch JH, Xu YB. High-throughput SNP genotyping with the GoldenGate assay in maize. Mol. Breed. 2009;25:441–451.

102. Emanuelli F, Battilana J, Costantini L, Le Cunff L, Boursiquot JM, This P, Grando MS. A candidate gene association study on muscat flavor in grapevine (*Vitis vinifera* L.) BMC Plant Biol. 2010;10:241.

103. Deulvot C, Charrel H, Marty A, Jacquin F, Donnadieu C, Lejeune-Hénaut I, Burstin J, Aubert G. Highly-multiplexed SNP genotyping for genetic mapping and germplasm diversity studies in pea. BMC Genomics. 2011;11:468.

104. Neumann K, Kobiljski B, Dencic S, Varshney RK, Borner A. Genome-wide association mapping: a case study in bread wheat (*Triticum aestivum* L.) Mol. Breed. 2011;27:37–58.

105. Chao S, Dubcovsky J, Luo MC, Baenziger SP, Matnyazov R, Clark DR, Talbert LE, Anderson JA, Dreisigacker S, Glover K, Chen J, Campbell K, Bruckner PL, Rudd JC, Haley S, Carver BF, Perry S, Sorrells ME, Akhunov ED. Population- and genome-specific patterns of linkage disequilibrium and SNP variation in spring and winter wheat (*Triticum aestivum* L.) BMC Genomics. 2011;11:727.

106. Beaulieu J, Doerksen T, Boyle B, Clément S, Deslauriers M, Beauseigle S, Poulin P, Lenz P, Caron S, Rigault P, Bicho P, Bousquet J, Mackay J. Association genetics of wood physical traits in the conifer white spruce and relationships with gene expression. Genetics. 2011;188:197–214.

107. Edenberg HJ, Yunlong L. Laboratory methods for high-throughput genotyping. Cold Spring Harb. Protoc. 2009;16:183–193.

108. Appleby N, Edwards D, Batley J. New technologies for ultra-high throughput genotyping in plants.Methods Mol. Biol. 2009;513:19–39.

109. Lin CH, Yeakley JM, McDaniel TK, Shen R. Medium- to high-throughput SNP genotyping using VeraCode microbeads. Methods Mol. Biol. 2009;496:129–142.

110. Gabriel S, Ziaugra L, Tabbaa D. SNP genotyping using the Sequenom MassARRAY iPLEX platform.Curr. Protocols Human Genet. 2009. pp. 2.12.1–2.12.18.

111. Eckert AJ, van Heerwaarden J, Wegrzyn JL, Nelson CD, Ross-Ibarra J, Gonzalez-Martinez SC, Neale DB. Patterns of population structure and environmental associations to aridity across the range of loblolly pine (*Pinus taeda* L., Pinaceae) Genetics. 2010;185:969–982.

112. Eckert AJ, Pande B, Ersoz ES, Wright MH, Rashbrook VK, Nicolet CM, Neale DB. High-throughput genotyping and mapping of single nucleotide polymorphisms in loblolly pine (*Pinus taeda* L.) Tree Genet. Genomes. 2009;5:225–234.

113. Liu S, Chen HD, Makarevitch I, Shirmer R, Emrich SJ, Dietrich CR, Barbazuk WB, Springer NM, Schnable PS. High-throughput genetic mapping of mutants via quantitative single nucleotide polymorphism typing. Genetics. 2010;184:19–26.

114. Neves L, Mamani E, Alfenas A, Kirst M, Grattapaglia D. A high-density transcript linkage map with 1,845 expressed genes

positioned by microarray-based Single Feature Polymorphisms (SFP) in *Eucalyptus*.BMC Genomics. 2011;12:189.

115. SatoK,NankakuN,TakedaK.Ahigh-densitytranscriptlinkagemapof barley derived from a single population. Heredity. 2009;103:110–117.

116. Chutimanitsakun Y, Nipper RW, Cuesta-Marcos A, Cistue L, Corey A, Filichkina T, Johnson EA, Hayes PM. Construction and application for QTL analysis of a Restriction Site Associated DNA (RAD) linkage map in barley. BMC Genomics. 2011;12:4.

117. Celton JM, Christoffels A, Sargent DJ, Xu XM, Rees DJG. Genome-wide SNP identification by high-throughput sequencing and selective mapping allows sequence assembly positioning using a framework genetic linkage map. BMC Biol. 2010;8:155.

118. Huang XH, Feng Q, Qian Q, Zhao Q, Wang L, Wang AH, Guan JP, Fan DL, Weng QJ, Huang T, Dong GJ, Sang T, Han B. High-throughput genotyping by whole-genome resequencing. Genome Res.2009;19:1068–1076.

119. Xu JJ, Zhao QA, Du PN, Xu CW, Wang BH, Feng Q, Liu QQ, Tang SZ, Gu MH, Han B, Liang GH. Developing high throughput genotyped chromosome segment substitution lines based on population whole-genome re-sequencing in rice (*Oryza sativa* L.) BMC Genomics. 2010;11:656.

120. XieWB, Feng Q, Yu HH, Huang XH, Zhao QA, XingYZ, Yu SB, Han B, Zhang QF. Parent-independent genotyping for constructing an ultrahigh-density linkage map based on population sequencing. Proc. Natl. Acad. Sci. USA. 2010;107:10578–10583.

121. Boerma R, Wilson R, Ready E. Soybean genomics research program strategic plan implementing research to meet 2012–2016 strategic milestones. Plant Genome. 2011;4:1–11.

122. Gore MA, Chia JM, Elshire RJ, Sun Q, Ersoz ES, Hurwitz BL, Peiffer JA, McMullen MD, Grills GS, Ross-Ibarra J, Ware DH, Buckler ES. A first-generation haplotype map of maize. Science. 2009;326:1115–1117.

123. Branca A, Paape T, Briskine R, Zhou P, Wang S, Denny R, Mudge J, Bharti AK, Farmer A, May GD, Tiffin PL, Young ND. The *Medicago truncatula* HapMap project: deep coverage sequencing of 30 inbred lines using Illumina's Solexa technology. Plant & Animal Genomes XVIII Conference, San Diego, CA.2010;P417

124. McNally KL, Childs KL, Bohnert R, Davidson RM, Zhao K, Ulat VJ, Zeller G, Clark RM, Hoen DR, Bureau TE, Stokowski R, Ballinger DG, Frazer KA, Cox DR, Padhukasahasram B, Bustamante CD, Weigel D, Mackill DJ, Bruskiewich RM, Rätsch G, Buell CR, Leung H, Leach JE. Genome wide SNP variation reveals relationships among landraces and modern varieties of rice. Proc. Natl. Acad. Sci. USA.2009;106:12273–12278.

125. Yamamoto T, Nagasaki H, Yonemaru J, Ebana K, Nakajima M, Shibaya T, Yano M. Definition of the pedigree haplotypes of closely related rice cultivars by means of genome-wide discovery of single-nucleotide polymorphisms. BMC Genomics. 2010;11:267. [PMC free article] [PubMed]

126. Finkel E. Imaging. With ‹phenomics,› plant scientists hope to shift breeding into overdrive. Science.2009;325:380–381.

127. Hyten DL, Smith JR, Frederick RD, Tucker ML, ong Q, Cregan PB. Bulked segregant analysis using the GoldenGate assay to locate the *Rpp3* locus that confers resistance to soybean rust in soybean. Crop Sci.2009;49:265–271.

128. Kang HX, Weng YQ, Yang YH, Zhang ZH, Zhang SP, Mao ZC, Cheng GH, Gu XF, Huang SW, Xie BY. Fine genetic mapping localizes cucumber scab resistance gene *Ccu* into an R gene cluster. Theor. Appl. Genet. 2011;122:795–803.

129. Yu HH, Xie WB, Wang J, Xing YZ, Xu CG, Li XH, Xiao JH, Zhang QF. Gains in QTL detection using an ultra-high density SNP map based on population sequencing relative to traditional RFLP/SSR markers. PLoS One, 2011;6(3):e17595.

130. Bräutigam A, Gowik U. What can next generation do for you. Plant Biol. 2010;12:831–841.

131. Buckler ES, Holland JB, Bradbury PJ, Acharya CB, Brown PJ, Browne C, Ersoz E, Flint-Garcia S, Garcia A, Glaubitz JC, Goodman MM, Harjes C, Guill K, Kroon DE, Larsson S, Lepak NK, Li HH, Mitchell SE, Pressoir G, Peiffer JA, Rosas MO, Rocheford TR, Romay MC, Romero S, Salvo S, Villeda HS, da Silva HS, Sun Q, Tian F, Upadyayula N, Ware D, Yates H, Yu JM, Zhang ZW, Kresovich S, McMullen MD. The genetic architecture of maize flowering time. Science. 2009;325:714–718.

132. Andersen JR, Zein I, Wenzel G, Darnhofer B, Eder J, Ouzunova M, Lubberstedt T. Characterization of phenylpropanoid pathway

genes within European maize (*Zea mays* L.) inbreds. BMC Plant Biol. 2008;8:2.

133. Harjes CE, Rocheford TR, Bai L, Brutnell TP, Kandianis CB, Sowinski SG, Stapleton AE, Vallabhaneni R, Williams M, Wurtzel ET, Yan JB, Buckle ES. Natural genetic variation in lycopene epsilon cyclase tapped for maize biofortification. Science. 2008;319:330–333.
134. Zheng P, Allen WB, Roesler K, Williams ME, Zhang S, Li J, Glassman K, Ranch J, Nubel D, Solawetz W, Bhattramakki D, Llaca V, Deschamps S, Zhong GY, Tarczynski MC, Shen B. A phenylalanine in DGAT is a key determinant of oil content and composition in maize. Nat. Genet.2008;40:367–372.
135. Manicacci D, Camus-Kulandaivelu L, Fourmann M, Arar C, Barrault S, Rousselet A, Feminias N, Consoli L, Frances L, Mechin V, Murigneux A, Prioul JL, Charcosset A, Damerval C. Epistatic interactions between Opaque2 transcriptional activator and its target gene CyPPDK1 control kernel trait variation in maize. Plant Physiol. 2009;150:506–520.
136. Atwell S, Huang YS, Vilhjalmsson BJ, Willems G, Horton M, Li Y, Meng DZ, Platt A, Tarone AM, Hu TT, Jiang R, Muliyati NW, Zhang X, Amer MA, Baxter I, Brachi B, Chory J, Dean C, Debieu M, de Meaux J, Ecker JR, Faure N, Kniskern JM, Jones JDG, Michael T, Nemri A, Roux F, Salt DE, Tang CL, Todesco M, Traw MB, Weigel D, Marjoram P, Borevitz JO, Bergelson J, Nordborg M. Genome-wide association study of 107 phenotypes in Arabidopsis thaliana inbred lines. Nature. 2010;465:627–631.
137. Simko I, Pechenick DA, McHale LK, Truco LJ, Ochoa OE, Michelmore RW, Scheffler BE. Association mapping and marker-assisted selection of the lettuce dieback resistance gene *Tvr1*. BMC Plant Biol. 2009;9:135–151.
138. Andersen JR, Lübberstedt T. Functional markers in plants. Trends Plant Sci. 2003;8:554–560.
139. Hospital F, Chevalet C, Mulsant P. Using m,02arkers in gene introgression breeding programs.Genetics. 1992;132:1199–1210. [PMC free article] [PubMed]
140. Gopalakrishnan S, Sharma RK, Anand Rajkumar KA, Joseph M, Singh VP, Bhat KV, Singh NK, Mohapatra T. Integrating marker assisted background analysis with foreground selection for

identification of superior bacterial blight resistant recombinants in Basmati rice. Plant Breed. 2008;127:131–139.

141. Cao Z, Tian F, Wang N, Jiang C, Lin B, Xia W, Shi J, Long Y, Zhang C, Meng J. Analysis of QTLs for erucic acid and oil content in seeds on A8 chromosome and the linkage drag between the alleles for the two traits in *Brassica napus*. J. Genet. Genomics. 2010;37:231–240.

142. Huang N, Angeles ER, Domingo J, Magpantay G, Singh S, Zhang G, Kumaravadivel N, Bennett J, Khush GS. Pyramiding of bacterial blight resistance genes in rice: marker-assisted selection using RFLP and PCR. Theor. Appl. Genet. 1997;95:313–320.

143. Ishii T, Yonezawa K. Optimization of the marker-based procedures for pyramiding genes from multiple donor lines: II. strategies for selecting the objective homozygous plant. Crop Sci. 2007;47:1878–1886.

144. Wei X, Liu L, Xu J, Jiang L, Zhang W, Wang J, Zhai H, Wan J. Breeding strategies for optimum heading date using genotypic information in rice. Mol. Breed. 2010;25:287–298.

145. Lü HY, Liu XF, Wei SP, Zhang YM. Epistatic association mapping in homozygous crop cultivars.PLoS One . 2011;6(3):e17773.

146. Jannink JL, Lorenz AJ, Iwata H. Genomic selection in plant breeding: from theory to practice. Brief. Funct. Genomics . 2010;9:166–177.

147. Meuwissen THE, Hayes BJ, Goddard ME. Prediction of total genetic value using genome-wide dense marker maps. Genetics. 2001;157:1819–1829.

148. Heffner EL, Sorrells ME, Jannink JL. Genomic selection for crop improvement. Crop Sci. 2009;49:1–12.

149. Bernardo R. Genomewide selection for rapid introgression of exotic germplasm in maize. Crop Sci.2009;49:419–425.

150. Bernardo R, Yu J. Prospects for genomewide selection for quantitative traits in maize. Crop Sci.2007;47:1082–1090.

151. Lorenzana RE, Bernardo R. Accuracy of genotypic value predictions for marker-based selection in biparental plant populations. Theor. Appl. Genet. 2009;120:151–161.

152. Bernardo R. Molecular markers and selection for complex traits in plants: learning from the last 20 years. Crop Sci. 2008;48:1649–1664.

153. Araus JL, Slafer GA, Royo C, Serret MD. Breeding for yield potential and stress adaptation in cereals.Crit. Rev. Plant Sci. 2008;27:377–412.

Citations

CHAPTER 1

Yunbi Xu, Debra J. Skinner, Huixia Wu, et al., "Advances in Maize Genomics and Their Value for Enhancing Genetic Gains from Breeding," International Journal of Plant Genomics, vol. 2009, Article ID 957602, 30 pages, 2009. doi:10.1155/2009/957602.

CHAPTER 2

H. Hartings, M. Fracassetti and M. Motto (2012). Genetic Enhancement of Grain Quality-Related Traits in Maize, Transgenic Plants - Advances and Limitations, PhD. Yelda Ozden Çiftçi (Ed.), ISBN: 978-953-51-0181-9, InTech, DOI: 10.5772/32758.

CHAPTER 3

W. Brien Henry, Gary L. Windham and Michael H. Blanco, Evaluation of maize germplasm for resistance to aflatoxin accumulation, doi: 10.3390/agronomy2010028

CHAPTER 4

Robert L. Brown, Deepak Bhatnagar, Thomas E. Cleveland, Zhi-Yuan Chen and Abebe Menkir (2013). Development of Maize Host Resistance to Aflatoxigenic Fungi, Aflatoxins - Recent Advances and Future Prospects, Prof. Mehdi Razzaghi-Abyaneh (Ed.), ISBN: 978-953-51-0904-4, InTech, DOI: 10.5772/54654.

CHAPTER 5

Haiping Ding, Cheng Qin, Xirong Luo, Lujiang Li, Zhe Chen, Hongjun Liu, Jian Gao,Haijian Lin,Yaou Shen, Maojun Zhao, Thomas Lübberstedt, Zhiming Zhang and Guangtang Pan; Heterosis in Early Maize Ear Inflorescence Development: A Genome-Wide Transcription Analysis for Two Maize Inbred Lines and Their Hybrid; doi: 10.3390/ijms150813892.

CHAPTER 6

Meizhen Wang, Chen Liu, Shixue Li,Dengyun Zhu, Qian Zhao, and Jingjuan Yu; Improved Nutritive Quality and Salt Resistance in Transgenic Maize by Simultaneously Overexpression of a Natural Lysine-Rich Protein Gene, SBgLR, and an ERF Transcription Factor Gene, TSRF1; doi:10.3390/ijms14059459

CHAPTER 7

A.M. Pérez-de-Castro, S. Vilanova, J. Cañizares, L. Pascual, J.M. Blanca, M.J. Díez, J. Prohens, and B. Picó, Application of Genomic Tools in Plant Breeding, doi: 10.2174/138920212800543084.

Index

A

B

C

D

E

F

G

H

I

K

L

M

U

Z